APPLIED HEAT TRANSFER
(With Worked Examples)

Volume II
Heat Convection in Fluids

B. N. Nnolim,

BSc(Lon), A. C. G.I., MSc (Lon), D. I. C., FNSChE, FNSE
Formerly of the Department of Chemical Engineering,
Nnamdi Azikiwe University, Awka,
Anambra State, Nigeria

First Published August, 1998
By
CECTA (NIG) LIMITED
159 Chime Avenue, New Haven.
P. O. Box 1306, Enugu.
Enugu State, Nigeria

This edition published July, 2010
ISBN 978-1-906914-22-6

Other Books by B. N. Nnolim

- APPLIED HEAT TRANSFER
  (With Worked Examples),
  Volume I: Conduction of Heat in Solids
  ISBN 978-1-906914-21-9

- FUNDAMENTALS OF MASS TRANSFER
  ISBN 978-1-906914-01-1

Ben Nnolim Books,
7 Sandway Path,
St. Mary Cray, Orpington, Kent.
BR5 3TS, UK
benedictnnolim@aol.com

Dedicated to

Professor R. W. H. Sargent,

BSc (Lon), PhD (Lon), F.C.G.I., F. I. C., C. Eng

Courtauld's Professor of Chemical Engineering,
Imperial College of Science, Technology & Medicine,
London SW7, UK

"Who can walk with kings
And not lose the common touch"

PREFACE TO THE FIRST EDITION

At the tertiary education level, the intention of study is to be initiated into the treasure house of knowledge of the human race and to prepare oneself to participate actively and creatively in the preservation and extension of this sum total of knowledge in a chosen discipline or sets of disciplines. The primary and secondary education one received is suddenly seen and appreciated in its true perspectives of creating awareness and sensitivity to natural and other phenomena and of preparing one to proceed in life from there.

Rote learning, form teaching and learning, combined with the current craze for '"hand-outs and examination malpractices, generally, in vogue today in Nigeria, distort and disorient this process. The joy of discernment and sudden illumination is lost. So also is the joy of reaping the fruits of one's study labour. For the teacher, the whole process seems to be just a waste of time while for the student, too late, he or she discovers that he or she has just cheated himself or herself. Every effort to better himself or herself, by taking post graduate and higher degree courses continue to be uphill tasks, demanding more cheating, more rote learning, etc. from such students.

The approach in this book, APPLIED HEAT TRANSFER, of which Volume II is HEAT CONVECTION IN FLUIDS, is to bring down to the level of students, without losing any breadth or depth (rigour), the various subject matters which are, customarily, presented, by both local and foreign authors, with an European/ American mindset. It is the belief of the author that a thorough grounding in fundamentals, gradually leading into professional, academic and industrial applications, in one book, fosters better understanding and creativity and hence usefulness in the subject. Serious students should, therefore, find it very useful.

A lot of material in this book owes its origin to many sources but I must acknowledge, specially, my indebtedness to the lectures in heat transfer, at Imperial College, London, in the 60s, by Prof. R. W. H. Sargent. The uniqueness and usefulness of these lectures lay, not only in their content, but mainly in an approach which related theory to every day, practical, industrial, research and development problems.

This book is organised into six chapters. Chapter 1 reviews fundamental concepts and definitions in heat convection, especially, the connection to, and relevance of, basic thermodynamics and fluid mechanics to heat convection. This gives the student perspective while preparing him or her for Chapter 2, which attempts to summarise the mathematical approaches that have been employed in tackling the theoretical and fundamental formulations of the heat convection process. Chapters 3 and 4 deal with the fact that the study of heat convection must be useful and, therefore, zero in on a practical concept, the heat transfer coefficient, which has been found, extremely, useful in the design and analysis of heat exchange devices. Chapter 5 deals, in some detail, with the theory and design of heat exchangers, which are, with pumps, the work horses of industry. The subject matter of Chapter 6, Optimisation of Heat Exchangers, is, often, treated superficially in standard text and only usefully in advanced texts. An attempt has now been made, in this chapter, to provide sound, modern knowledge in the subject.

Illustrative problems, as worked examples, are given at the end of every chapter and, while not covering every thing, are intended to extend the understanding of the student in the areas treated in the body of the text.

I thank my wife, Dorothy, and my children, Neme, Chukwuma, Chiemedinam and Uche for their forbearance and support throughout the writing and production of this book.

I thank our Blessed Virgin Mary, The Immaculate Conception, for her intercession and our God for His love, kindness and protection

Ben Nnolim
August 20, 1998.

PREFACE TO THIS EDITION

Engineering in Africa, south of the Sahara, continues to be regarded as, depending on the part of Africa you are in, a properly whiteman's profession at which Africans will never be good, a challenging educational activity meant only for a special intellectual elite, or a prestigious profession with which go bumper salaries and social power. Rarely is it seen, by African engineers, as an opportunity to improve the welfare or living conditions of their environment or society.

In spite of the thousands of African engineers, trained in the best and worst universities and other higher institutions in every conceivable country in the world, an uncomfortably large percentage of Africa's engineering problems are still being solved by expatriate engineers and technologists. Turnkey and grassroots projects executed by these expatriates are unable to be managed and maintained properly or successfully by African engineers as it takes only a few years before the expatriates have to be called in again to repair, upgrade or replace these projects.

In most of Africa, the strategic resources are, still, agricultural and mineral materials while the transforming resources are, still, human and animal labour. To solve their technical, economic and social problems, they need the strategic resource of money capital and the transforming resource of generated power. The Western world, controlling the world and its economy, has passed through these stages and now has knowledge as its strategic resource and data processing as its transforming resource. Africa needs, therefore, to operate at the level in which, while retaining its strategic agricultural and mineral resources and their transforming resource of human labour, it indigenises control of the strategic resources of money capital and knowledge and their transforming resources of generated power and data processing.

Commercial and large scale processing of materials depend on proper understanding and harnessing of the principles of momentum, heat and mass transfer and thermodynamics/chemical reaction. Almost all heat transfer equipment for domestic and commercial use in Africa is imported from Europe, the Americas or the Far East. There appears to be little interest or ability among local engineers to develop, design and manufacture heat transfer equipment that can be used in indigenous technologies for which there are no Western analogs and which, not unsurprisingly, still account for the provision of a majority of the human needs of the average African.

The first edition of this book attempted to address this issue by trying to bring down from their intellectual heights, engineering heat transfer concepts, ideas and methods in convective heat transfer in the hope that students would readily understand them and do not get intimidated and resort to rote learning in order to get along. The aim was to make these topics ordinary enough for the student to understand that they are necessary tools for solving every day convective heat transfer problems and not new, esoteric and mystifying concepts and jargon for mystifying and impressing one's contemporaries. The hope was that, with full and fundamental understanding of these concepts, the African engineer can apply them to the improvement, to the mass production stage and availability, of their indigenous technologies. These objectives are still aimed at in this edition.

The first edition was, also, full of errors most of which have now been corrected in this edition. There has also been a rearrangement of the chapters, some completely rewritten, with most of Chapters 1 and 2 in the first edition going to the Appendices in this edition. New information have also been added based on new developments but the emphasis continues to be the desire to ground the students in fundamentals from which they can progress to applying these to their local environment as well as interface with modern concepts, ideas, products and processes in the subject.

No work of this kind can be without error. Every effort has been made to avoid them. I apologise, however, for those inevitable errors which seem to occur no matter how hard you try to prevent them from occurring.

I continue to thank Almighty God who has kept me alive inspite of several life threatening ailments and Our Blessed Virgin Mary, the Immaculate Conception, and all the angels and saints in heaven whose intercessions have never failed.

Benedict Nnolim,
July 20, 2010

Table of Contents

CHAPTER ONE
FUNDAMENTAL CONCEPTS AND DEFINITIONS

1.0: Heat

Heat, also known as thermal energy, is energy in transition between a system and its surroundings as a result of a temperature difference. This transition may occur by a transport or transfer mechanism. Transport occurs within a phase, and is by means of bulk motion while transfer takes place across an interface or boundary and as a result of molecular motion. Unless work is done for the purpose, heat transfer always takes place from a region of high temperature to one of lower temperature.

Heat energy may be transferred by conduction, convection, radiation or by any combination of these. *Conduction* is the transfer or transport of heat from one part of a body to another, in physical contact with it, without appreciable displacement of the particles of the body. It is, usually, the result of an interchange of kinetic energy between molecules. *Convection* is the transport of heat from one point to another within a fluid (gas or liquid) by the mixing of one portion of the fluid with another. Often it occurs by the combination of heat conduction and circulation or movement of hot particles in bulk. *Radiation* occurs when energy is transferred by electromagnetic waves from a body at high temperature to one at a lower temperature which is not in contact with it.

Convective heat transfer is important, commercially, because of its usefulness as an energy utilisation and/or recovery route in many domestic appliances and industrial processes. Through convective heat transfer, fluid streams at high temperature, undergoing combustion or other energy release processes and thus containing large quantities of heat energy, transfer their energy to other processes and systems either directly or indirectly by means of devices such as radiators, boilers, heat exchangers, chemical or other reactors, etc.

1.1: Energy Balance in a System in Heat Transfer

Since the energy of the universe is constant, an energy balance must be the first consideration in any application. In any given problem, the universe is taken to be the control volume or surface, with flexible boundaries, which define the physical system under consideration. The relevant forms of energy, for this energy balance, are the internal, potential, kinetic, work, thermal, radiant or solar, latent, electric, electromagnetic, mass or nuclear and sonic (a form of kinetic) energies. These are, usually, combined in a general energy balance, in the chosen control volume or surface, using the principle of conservation of energy. The mathematical result of this combined energy balance is known as the general energy equation.

The control volumes or surfaces, we often deal with, are for the so called *PVT* systems, that is, systems whose behavior is determined, generally, by the pressure, volume and temperature of themselves and/or of their surroundings. The energy which these *PVT* systems have to manage can be generated by fossil fuels, electromagnetic induction or by solar or nuclear energy.

For such control volumes or surfaces, the most general situation is represented as one in which all forms of energy enter the control volume or surface, external work is done and unused energy (according to the second law of thermodynamics) leaves the system. In such a system or control volume or surface, some energy may be generated within the system and some energy may be accumulated also within the system. Such a system is, generally, represented as shown in Fig. 1.1.

Figure 1.1: General Energy Balance Diagram

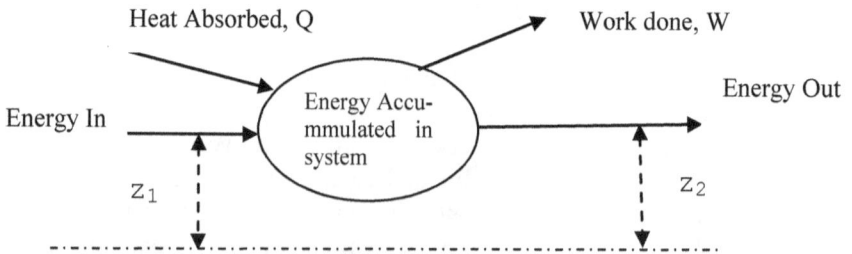

According to the principle of conservation of energy, which is also a statement of the first law of thermodynamics, the energy balance for any system would be

$$ENERGY\ GOING\ IN + ENERGY\ GENERATED\ OR\ CONSUMED$$
$$= ENERGY\ GOING\ OUT + ENERGY\ ACCUMULATED \qquad (1.0)$$

For an open PVT system, which can exchange both mass and energy with its surroundings, this principle, of conservation of energy, can be

2

expressed, on per unit mass basis, as a steady state general energy equation given by

$$U_1 + P_1V_1 + \frac{u_1^2}{2} + g\,z_1 + Q - W_F = U_2 + P_2V_2 + \frac{u_2^2}{2} + g\,z_2 \qquad (1.1)$$

where

$\quad\quad\quad$ U = internal energy, J/kg
$\quad\quad\quad$ P = pressure, N/m^2 or Pascals
$\quad\quad\quad$ V = volume per unit mass, m^3/kg
$\quad\quad\quad$ u = velocity, m/s
$\quad\quad\quad$ z = height of fluid above a chosen datum level, m
$\quad\quad\quad$ Q = heat absorbed within the control volume, J/kg
$\quad\quad\quad$ W$_F$ = work done against friction within the control volume, J/kg
$\quad\quad\quad$ g = the acceleration due to gravity, 9.81m^2/s

Subscripts 1 and 2 indicate the initial and final states in which these energy terms are evaluated. Since the enthalpy, H, is given by

$$H = U + PV \qquad (1.2)$$

the steady state general energy equation may be rewritten as:

$$H_1 + \frac{u_1^2}{2} + g\,z_1 + Q - W_F = H_2 + \frac{u_2^2}{2} + g\,z_2 \qquad (1.3)$$

In chemical processes, the kinetic and potential energy terms are, usually, small compared to the heat and work terms, so that:

$$H_2 - H_1 = Q - W_F \qquad (1.4)$$

Often too, the work term is negligible, so that:

$$Q = H_2 - H_1 \qquad (1.4a)$$

where

$$Q = Q_{gen} + Q_{add} \qquad (1.5)$$

and Q_{gen} = heat generated in the system and Q_{add} = heat added to the system, either of which can be positive or negative.

For a closed system, which exchanges energy only, but not mass, with its surroundings, the first law of thermodynamics states that the change in internal energy is equal to the heat absorbed minus the work done.

$$dU = dQ - PdV \qquad (1.6)$$

The differential form of equation (1.3) is

$$dH + u\,du + g\,dz = dQ - dW_F \qquad (1.7)$$

The differential form of equation (1.2) is

$$dH = dU + PdV + VdP \qquad (1.8)$$

Substituting equation (1.8) in equation (1.7)

$$dU + PdV + VdP + u\,du + g\,dz = dQ - dW_F \qquad (1.9)$$

3

Comparing equation (1.6) with equation (1.9), it can be seen that

$$VdP + u\,du + g\,dz = -dW_F \qquad (1.10)$$

Equation (1.10) is known as the Euler's (pronounced *oiler's*) equation or the mechanical energy equation. It shows that the mechanical energy of a real fluid, in motion, decreases as a result of fluid friction and turbulence. When dW_F is zero, equation (1.10) reduces to the Bernoulli's equation for the flow of a frictionless fluid, for which the mechanical energy is constant.

For the flow of a real fluid, in a system with fixed boundaries, there is no slip at the solid boundary and, hence, no work can be done at the fixed surface, even though shear stress, due to viscosity, may be present. For such cases, $W_F = 0$ and equation (1.3) becomes

$$H_1 + \frac{u_1^2}{2} + g\,z_1 + Q = H_2 + \frac{u_2^2}{2} + g\,z_2 \qquad (1.11)$$

That work cannot be done at a fixed surface, by a real fluid, does not imply that heat cannot be generated internally in the fluid. Thus, by the first law of thermodynamics, for a closed system, we can say that:

$$dU = dQ_{gen} + dQ_{add} - PdV \qquad (1.12)$$

where dQ_{add} = differential quantity of heat absorbed and dQ_{gen} = differential quantity of heat generated, internally, within the system by fluid friction and turbulence.

For flow of a real fluid past moving boundaries, such as the blades of a turbine or compressor, shaft work, W_S, can be expected to be done by the fluid. Then, the energy equation (1.3) becomes:

$$H_1 + \frac{u_1^2}{2} + g\,z_1 + Q - W_s = H_2 + \frac{u_2^2}{2} + g\,z_2 \qquad (1.13)$$

with the following special cases:

a). When no external work is done, there is no change in velocity. This is typical of flow through heat exchangers. That is $W_s = 0,\ u_1 = u_2,\ z_1 = z_2$ and

$$Q = H_2 - H_1 \qquad (1.14)$$

b). There is no transfer of heat nor significant change in velocity as in flow through a compressor. That is $Q = 0,\ u_1 = u_2,\ z_1 = z_2$ and

$$W_s = H_2 - H_1 \qquad (1.15)$$

c). There is no external work nor transfer of heat such as occur in flow through a nozzle. That is $Q = 0,\ W_s = 0,\ z_1 = z_2$ and

4

$$\frac{u_1^2}{2} - \frac{u_2^2}{2} = H_2 - H_1 \qquad (1.16)$$

d). No external work is done, there is no transfer of heat nor significant change in velocity, such as occur in a throttling process. That is $W_s = 0$, $Q = 0$, $u_1 = u_2$, $z_1 = z_2$ and

$$H_2 = H_1 \qquad (1.17)$$

1.2: The Equations of Motion and of Heat Transfer

Now that we know about the general energy balance in a PVT system, another fundamental concept that needs to be described and understood is the motion of heated fluids in such systems. Since heat transfer by convection involves fluid motion, it is important to be familiar with the equations governing fluid motion. These are the Cauchy equations, the continuity equations, the Navier - Stokes and the energy equations.

Another equation that is involved in the description of fluid motion is the mass transfer equation, sometimes referred to as the Ficks second law of mass transfer.

All these equations are, often stated in vector notation because it is the most concise form of stating them and also the form in which their, sometimes, complex mathematics can be handled conveniently. They are all based on these very fundamental assumptions, namely

1. that the fluid, at the scale of interest, is a continuum (a continuous substance, not made up of discrete particles)
2. that all variables involved such as pressure, velocity, temperature, density, etc, are differentiable, at least within the scale of analysis of the variables.
3. that Eulerian analysis, that is, observing changes in velocity or acceleration from constant points in space, is in use

1.2.1: The Cauchy Equation of Motion

The Cauchy equation of motion (Cauchy momentum equation) describes the motion of any material, fluid or otherwise, by substituting the expressions for the acceleration, the body forces, the pressure gradient forces and the viscous forces in the material, into the expression for Newton's second law of motion. In vector notation, it can be expressed as

$$\rho \frac{D\bar{v}}{Dt} = \bar{f} + \nabla \bullet \sigma_{ij} \qquad (1.18)$$

where v with the overbar is the velocity vector, t, the time, ρ, the material density, f with the overbar, the vector value of the body force, and σ the viscous stress tensor (also known as a rank two symmetric tensor) given by its covariant components.

$$\sigma_{ij} = \begin{pmatrix} \sigma_{xx} & \tau_{xy} & \tau_{xz} \\ \tau_{yx} & \sigma_{yy} & \tau_{yz} \\ \tau_{zx} & \tau_{zy} & \sigma_{zz} \end{pmatrix} = - \begin{pmatrix} P & 0 & 0 \\ 0 & P & 0 \\ 0 & 0 & P \end{pmatrix}$$

$$+ \begin{pmatrix} \sigma_{xx} + P & \tau_{xy} & \tau_{xz} \\ \tau_{yx} & \sigma_{yy} + P & \tau_{yz} \\ \tau_{zx} & \tau_{zy} & \sigma_{zz} + P \end{pmatrix} = - PI + \overline{T_{ij}} \qquad (1.19)$$

I is the identity matrix. P is the absolute pressure and is given by

$$P = \frac{1}{3}\left(\sigma_{xx} + \sigma_{yy} + \sigma_{zz}\right) \qquad (1.20)$$

The body force is the gravity force component which acts only in the z-direction but not in the x and y-directions. It is given as

$$\bar{f} = -\rho g \bar{k} \qquad (1.21)$$

where k with the overbar is the second of the three unit vector constants, i, j, k. Thus equation (1.18) becomes

$$\rho \frac{D\bar{v}}{Dt} = -\rho g \bar{k} - \nabla P + \nabla \bullet \overline{T_{ij}} \qquad (1.22)$$

The scalar form, in Cartesian co-ordinates, of equation (1.22) is

$$\rho \left[\frac{\partial u}{\partial t} + u\frac{\partial u}{\partial x} + v\frac{\partial u}{\partial y} + w\frac{\partial u}{\partial z} \right] = -\frac{\partial P}{\partial x} + \frac{\partial \tau_{xx}}{\partial x} + \frac{\partial \tau_{yx}}{\partial y} + \frac{\partial \tau_{zx}}{\partial z} \qquad (1.23a)$$

$$\rho \left[\frac{\partial v}{\partial t} + u\frac{\partial v}{\partial x} + v\frac{\partial v}{\partial y} + w\frac{\partial v}{\partial z} \right] = -\frac{\partial P}{\partial y} + \frac{\partial \tau_{xy}}{\partial x} + \frac{\partial \tau_{yy}}{\partial y} + \frac{\partial \tau_{zy}}{\partial z} \qquad (1.23b)$$

$$\rho \left[\frac{\partial w}{\partial t} + u\frac{\partial w}{\partial x} + v\frac{\partial w}{\partial y} + w\frac{\partial w}{\partial z} \right] = -\rho g - \frac{\partial P}{\partial z} + \frac{\partial \tau_{xz}}{\partial x} + \frac{\partial \tau_{yz}}{\partial y} + \frac{\partial \tau_{zz}}{\partial z} \qquad (1.23c)$$

In order to solve the Cauchy equations for any substance, be it rock, plastic, liquid or gas, we need to know the relationship (the constitutive relationship) between stress and strain in that substance. For fluids, the relationship is the mathematical one between the stress tensor, τ_{ij} and the strain rate tensor, ε_{ij}.

1.2.2: The Continuity Equation

This is derived from the mathematical statement of the law of conservation of mass applied to fluid flow. This law states, simply, that the rate of accumulation of mass inside a control surface is equal to the net rate of inflow of mass (inflow minus outflow) across the control surface plus what is created or consumed by sources or sinks inside the control volume. Note that this is another way of stating equation (1.0).

In vector notation, equation (1.0), on mass only basis, may be stated as

$$\frac{\partial}{\partial t}\oint_V \rho \, dV + \oint_S \rho \bar{v} \bullet n \, dS + \oint_V Q \, dV = 0 \tag{1.24}$$

Equation (1.24) is a statement of the Reynolds transport theorem where the intensive property is density, ρ, and v, with the overbar, is the velocity of the fluid and Q, in this case, represents the sources and sinks in the fluid.

Gauss's divergence theorem can be applied to the surface integral in equation (1.24) to convert all of its terms to volume integrals. Gauss's theorem states that

$$\oint_S \rho u \, n \, dS = \oint_V div \rho u \, dV = \oint_V \nabla \bullet (\rho u) \, dV \tag{1.25}$$

This enables us to arrange equation (1.24) as

$$\frac{\partial}{\partial t}\oint_V \rho \, dV + \oint_V div \rho u \, dV + \oint_V Q \, dV = 0 \tag{1.26}$$

Equation (1.26) can be rearranged by Leibniz's rule, again, as

$$\frac{\partial \rho}{\partial t} + div \rho u + Q = 0 \tag{1.27}$$

When mass is conserved, Q = 0 and

$$\frac{\partial \rho}{\partial t} + div \rho u = 0 \tag{1.28}$$

This is the mass continuity equation. It is amenable to the following simplifications:

For steady flow, $\frac{\partial}{\partial t} = 0$ and

$$div \rho u = 0 \tag{1.29}$$

For incompressible fluids, ρ is constant and $\dfrac{\partial \rho}{\partial t} = 0$, so that

$$divu = \nabla \bullet u = 0 \tag{1.30}$$

which is a statement of the conservation of volume. The continuity equation is always used together with the Navier-Stokes equation.

1.2.3. The Navier – Stokes Equations

The Navier-Stokes equations arise from being able to express the divergence of the stress tensor in the Cauchy's momentum equation (1.18) or (1.22) in terms of fluid velocities using the constitutive relationship for the particular fluid.

Newtonian Fluids

For incompressible Newtonian fluids, the constitutive relationship is found to be of the form

$$\tau_{ij} = 2\mu \varepsilon_{ij} \tag{1.31}$$

where

$$\varepsilon_{xx} = \frac{\partial u}{\partial x}, \ \varepsilon_{yy} = \frac{\partial v}{\partial y}, \ \varepsilon_{zz} = \frac{\partial w}{\partial z} \ and \ \varepsilon_{ij} = \frac{1}{2}\left(\frac{\partial u_i}{\partial x_j} + \frac{\partial u_j}{\partial x_i} \right) \tag{1.32}$$

Equation (1.31) has also been put into the concise form

$$\tau_{ij} = \mu\left(\frac{\partial u_i}{\partial x_j} + \frac{\partial u_j}{\partial x_i} \right) + \delta_{ij}\lambda\nabla \bullet \bar{v} \tag{1.33}$$

where δ_{ij} is the Kronecker delta function, μ and λ are constants which assume that stress and strain are linearly related. μ is known as the first coefficient of viscosity and λ as the second coefficient of viscosity. λ is related to bulk viscosity (viscous effects associated with volume change). It is said to be negligible in compressible flow and takes the approximate value

$$\lambda = \frac{2}{3}\mu \tag{1.34}$$

in nearly incompressible fluids. Using equation (1.33) in the conservation of momentum equation similar to equation (1.24) or equation (1.31) and (1.32) in the Cauchy's momentum equation (1.22), the Navier-Stokes equation is derived to be

$$\rho \frac{D\bar{v}}{Dt} = -\nabla P_x - \rho g \bar{k} + \mu \nabla^2 \bar{v} + \frac{\mu}{3} \nabla.div\bar{v} \qquad (1.35)$$

or

$$\rho \frac{D\bar{v}}{Dt} = -\nabla P_x - \rho g \bar{k} + \mu \nabla^2 \bar{v} \qquad (1.36)$$

since $\nabla \bullet \bar{v} = 0$ for incompressible flow. The scalar form of equation (1.36) is, in Cartesian co-ordinates,

$$\rho \left[\frac{\partial u}{\partial t} + u \frac{\partial u}{\partial x} + v \frac{\partial u}{\partial y} + w \frac{\partial u}{\partial z} \right] = -\frac{\partial P}{\partial x} + \mu \left[\frac{\partial^2 u}{\partial x^2} + \frac{\partial^2 u}{\partial y^2} + \frac{\partial^2 u}{\partial z^2} \right] \qquad (1.37a)$$

$$\rho \left[\frac{\partial v}{\partial t} + u \frac{\partial v}{\partial x} + v \frac{\partial v}{\partial y} + w \frac{\partial v}{\partial z} \right] = -\frac{\partial P}{\partial y} + \mu \left[\frac{\partial^2 v}{\partial x^2} + \frac{\partial^2 v}{\partial y^2} + \frac{\partial^2 v}{\partial z^2} \right] \qquad (1.37b)$$

$$\rho \left[\frac{\partial w}{\partial t} + u \frac{\partial w}{\partial x} + v \frac{\partial w}{\partial y} + w \frac{\partial w}{\partial z} \right] = -\rho g - \frac{\partial P}{\partial z} + \mu \left[\frac{\partial^2 w}{\partial x^2} + \frac{\partial^2 w}{\partial y^2} + \frac{\partial^2 w}{\partial z^2} \right] \qquad (1.37c)$$

The polar and spherical co-ordinate forms are listed in Appendix II.

Bingham Fluids

Bingham fluids, such as clay, toothpaste, lava, mud, etc, are capable of bearing shear stress up to a critical stress, τ_0, after which the fluids begin to move under shear. The constitutive relation is

$$\frac{\partial u}{\partial y} = \begin{vmatrix} 0, & \tau < \tau_0 \\ \frac{\tau - \tau_0}{\mu}, & \tau > \tau_0 \end{vmatrix} \qquad (1.38)$$

Power law Fluids

Known also as the Ostwald de Waele power law, it is an idealized mathematical model of the constitutive relation of non-Newtonian fluids which has the advantage of greater simplicity and tolerable accuracy compared to more exact but complicated models. It is stated as

$$\tau = K \left(\frac{\partial u}{\partial y} \right)^n \qquad (1.39)$$

where K is the flow consistency index with units of $Pa \cdot s^n$. It is useful for shear thinning fluids such as latex paint and shear thickening fluids such

9

as corn starch/water mixture.

To obtain the Navier-Stokes equations for these fluids similar to those obtained for Newtonian fluids, equations (1.38) and (1.39) have to be substituted in the appropriate form of the momentum balance equation such as (1.22)

1.2.4: The Energy Equation

In similar fashion to mass and momentum balances, the energy equation, for Newtonian fluids, is derived, from an energy balance, to be

$$\rho \frac{D\bar{h}}{Dt} = - \frac{DP}{Dt} + \nabla \bullet (k \nabla T) + \Phi \qquad (1.40)$$

where h is the enthalpy, T is absolute temperature and Φ is energy dissipated as a result of viscous effects and is given by

$$\Phi = 2\mu \left[\left(\frac{\partial u}{\partial x} \right)^2 + \left(\frac{\partial v}{\partial y} \right)^2 + \left(\frac{\partial w}{\partial z} \right)^2 \right] +$$

$$\mu \left[\left(\frac{\partial v}{\partial x} + \frac{\partial u}{\partial y} \right)^2 + \left(\frac{\partial w}{\partial y} + \frac{\partial v}{\partial z} \right)^2 + \left(\frac{\partial u}{\partial z} + \frac{\partial w}{\partial x} \right)^2 \right] \qquad (1.41)$$

An appropriate equation of state, most commonly the ideal gas equation of state, is also required to reflect, accurately, the relationships between pressure, volume and temperature changes. Note that enthalpy, $h = CpT$ for a system in constant total pressure. Note, also, that the pressure, in these equations, is the absolute, not the guage, pressure. The scalar form of equation (1.40), when Φ is negligible, is

$$\rho Cp \left[\frac{\partial T}{\partial t} + u \frac{\partial T}{\partial x} + v \frac{\partial T}{\partial y} + w \frac{\partial T}{\partial z} \right] = - \frac{\partial P}{\partial x} + \mu \left[\frac{\partial^2 T}{\partial x^2} + \frac{\partial^2 T}{\partial y^2} + \frac{\partial^2 T}{\partial z^2} \right] \qquad (1.42)$$

$$\rho Cp \left[\frac{\partial T}{\partial t} + u \frac{\partial T}{\partial x} + v \frac{\partial T}{\partial y} + w \frac{\partial T}{\partial z} \right] = - \frac{\partial P}{\partial y} + \mu \left[\frac{\partial^2 T}{\partial x^2} + \frac{\partial^2 T}{\partial y^2} + \frac{\partial^2 T}{\partial z^2} \right] \qquad (1.43)$$

$$\rho Cp \left[\frac{\partial T}{\partial t} + u \frac{\partial T}{\partial x} + v \frac{\partial T}{\partial y} + w \frac{\partial T}{\partial z} \right] = - \frac{\partial P}{\partial z} + \mu \left[\frac{\partial^2 T}{\partial x^2} + \frac{\partial^2 T}{\partial y^2} + \frac{\partial^2 T}{\partial z^2} \right] \qquad (1.44)$$

The polar and spherical co-ordinate forms are, also, listed in Appendix II.

1.3: Heat Convection

Heat convection has, already, been defined as the transport of heat from one point to another within a fluid (gas or liquid) by the mixing of one portion of the fluid with another. Often it occurs by the combination of heat conduction and transfer by circulation or movement of hot particles in bulk. Heat convection occurs only in liquids and gases and in solids only when they are in molten form.

There are two kinds of heat convection, namely, natural convection and forced convection. In *natural* convection, bulk motion of fluid is as a result of density differences (buoyancy) which arise from temperature differences, usually, in a gravity field. In *forced* convection, however, bulk motion of fluid is produced by mechanical means such as a pump or fan, etc.

1.3.1: Natural or Free Convection

In natural convection, fluid surrounding a heat source receives heat, becomes less dense and rises. The surrounding cooler fluid then moves to replace it. This cooler fluid is then heated and the process continues, forming convection currents, thereby transferring heat energy from the bottom of the convection cells to the top. The driving force for natural convection is buoyancy, a result of differences in fluid density. For natural convection to continue to take place, it is essential that there exists proper (rather than co-ordinate) acceleration.

Proper acceleration is that acceleration which arises from a resistance to gravity from an equivalent force. Co-ordinate acceleration is dependent on choice of coordinate systems and thus upon choice of observers. The equivalent force may arise from acceleration, centrifugal or Coriolis force. Natural convection, for example, essentially does not operate in free-fall (inertial) environments.

Common examples of natural convection include convection cells formed from air rising above sunlight-warmed land or water, in the rising plume of hot air from fires, oceanic currents, and sea-wind formation (where upward convection is also modified by Coriolis forces). In engineering applications, free convection is commonly visualized in the formation of microstructures during the cooling of molten metals, when fluid flows around heat dissipation fins, and in solar ponds.

A very common industrial application of natural convection is in the cooling of rooms or appliances using free air without the aid of fans. This happens all the time in the way architects design homes and on a small scale in the cooling of computer chips in desk top computers.

Because density differences imply volume differences, it is important to be able to evaluate such volume differences, usually, in terms of a volume expansion coefficient. The volume expansion coefficient, β, is, generally, defined as

$$\beta = \frac{1}{V}\frac{\partial V}{\partial T}\bigg|_P = \frac{1}{v}\frac{\partial v}{\partial T}\bigg|_P = -\frac{1}{\rho}\frac{\partial \rho}{\partial T}\bigg|_P \qquad (1.45)$$

where v is the specific volume. This parameter β is also known as the volume expansivity and has the units of degree K^{-1}.

For example, for an ideal gas, since PV = nRT

$$n = \frac{mass(m)}{molecular weight} \quad so\ that\ \frac{n}{V} = \rho = \frac{P}{RT} = molar density \qquad (1.46)$$

$$\frac{d\rho}{dT} = -\frac{P}{R}\cdot\frac{1}{T^2} \qquad (1.47)$$

Combining equations (1.45), (1.46) and (1.47)

$$\beta = -\frac{1}{\rho}\frac{d\rho}{dT} = -\left(\frac{RT}{P}\right)\left(-\frac{P}{R}\cdot\frac{1}{T^2}\right) = \frac{1}{T} \qquad (1.48)$$

For liquids, volume expansion, due to temperature changes, at constant total pressure, is assumed to be linear and the variation of density with temperature is found to be, generally,

$$\rho = \rho_0\left(1 + \beta\,\Delta T\right) \qquad (1.49)$$

ΔT is the temperature difference between the hot surface and the bulk fluid (K) and ρ_0 is the density at the temperature from which a temperature change is measured.

Mathematically, the criterion, which determines whether any system is in natural convection or not, is the Grashof number (Gr), which is a ratio of buoyancy to viscous forces, per unit volume. The bouyancy force per unit volume, F_b, can be seen from equation (1.49) to be

$$F_b = (\rho - \rho_0)g = \rho_0\,\beta\,\Delta T\,g \qquad (1.50)$$

This tells us that the Grashof number will then be

$$Gr = \frac{Bouyancy Force/Unit Volume}{Viscous Force/Unit Volume} = \frac{\rho_0 \, \beta \, g \, \Delta T}{\dfrac{\mu^2}{\rho_0 \, L^3}} = \frac{\rho_0^2 g \, \beta \, \Delta T \, L^3}{\mu^2} \qquad (1.51)$$

g is acceleration due to gravity, L is the characteristic length of the object and μ is the viscosity of the liquid.

A similar equation can be written for natural convection arising from a concentration gradient, ΔC, also known as thermo-solutal convection. In this case, a concentration of hot fluid diffuses into a cold fluid, in much the same way as ink poured into a container of water would, and diffuses to dye the entire space. In this case

$$Gr = \frac{g \, \beta \, \Delta C \, L^3}{\mu^2} \qquad (1.52)$$

Coming back to temperature driven natural convection, the relative magnitudes of the Grashof and Reynolds number determine which form of convection dominates.

If $\dfrac{Gr}{Re^2} \gg 1$, forced convection may be neglected, whereas if $\dfrac{Gr}{Re^2} \ll 1$ natural convection may be neglected. If $\dfrac{Gr}{Re^2} \approx 1$, both forced and natural convection need to be taken into account.

1.3.1.1: Equations of Fluid Motion in Natural Convection

The mass continuity equation [either (1.28), (1.29) or (1.30)] is still the same so that, for an incompressible fluid, the continuity equation is $\qquad\qquad divu = \nabla \bullet u = 0 \qquad\qquad (1.30)$
Equation (1.45) or (1.48) would suggest that

$$\rho = \rho_0 \, e^{-\beta(T - T_0)} \qquad (1.51)$$

where T_0 and ρ_0 are the temperature and density at some reference state. For small values of $\beta(T-T_0)$, equation (1.51) reduces to

$$\rho = \rho_0 \left[1 - \beta(T - T_0) + \frac{\beta^2(T - T_0)^2}{2!} - \frac{\beta^3(T - T_0)^3}{3!} + ... \right]$$

$$= \rho_0 \left[1 - \beta(T - T_0) \right] \qquad (1.52)$$

Since $\nabla P = \rho_0 g$, in free convection (hydrostatic pressure only), the momentum equation (1.36) becomes, in natural convection,

$$\rho\frac{D\bar{v}}{Dt} = -\rho_0\beta g\,(T - T_0) + \mu\nabla^2\bar{v} \tag{1.53}$$

1.3.2: Forced Convection

Forced convection occurs when the fluid is forced to flow over any surface by an external agent such as a fan or a pump which creates an artificially induced convection current. In this case, the continuity equation is still, for an incompressible fluid,

$$divu = \nabla \bullet u = 0 \tag{1.30}$$

The momentum equation is still

$$\rho\frac{D\bar{v}}{Dt} = -\nabla P_x - \rho g\bar{k} + \mu\nabla^2\bar{v} \tag{1.36}$$

1.3.3: Free and Forced Convection

When free and forced convection take place simultaneously, that is when $\frac{Gr}{Re^2} = 1$, $\nabla P = \rho_0 g + \nabla p$ (sum of hydrostatic and dynamic pressures), the momentum equation (1.36) becomes

$$\rho\frac{D\bar{v}}{Dt} = \nabla P - \rho_0\beta g\,(T - T_0) + \mu\nabla^2\bar{v} \tag{1.54}$$

1.4: Regimes of Fluid Flow in Heat Convection

Because heat convection in fluids is associated with fluid flow, common phenomena and characteristics of fluid flow such as laminar flow, boundary layers and turbulence must be considered in any analysis of heat convection in fluids.

1.4.1: Laminar Flow in Heat Convection

A real fluid has friction or viscosity both within the fluid and between the fluid and any solid boundary in contact with it. Because of this viscosity, the different layers of a real fluid in motion, and at different distances from the solid boundary, do not move at the same velocity as they would do in an inviscid fluid. If the solid boundary is stationary, the fluid velocity would vary from zero, at the boundary (the

14

no slip condition), to a maximum somewhere away from the boundary.

The fluid layers near the solid boundary are called boundary layers if their velocity gradients are not zero. Because the fluid layers near the solid boundary do not move as fast as those farther away from it, their behaviour become, in convective heat transfer, the major determinant of the rate of heat transfer across the solid boundary.

Thus, if there is a temperature difference between the solid boundary and the main body of the fluid, there would, also, be fluid layers, each at a different temperature, at varying distances from the solid wall. The boundary layers formed as a result of velocity differences are, thus, called hydrodynamic boundary layers while those formed as a result of temperature differences are known as thermal boundary layers.

Figure 1.2: Velocity Profiles in Viscid and Inviscid Fluid Flow

Velocity in Layers of Inviscid Fluid   **Velocity in Layers of Real Fluid**

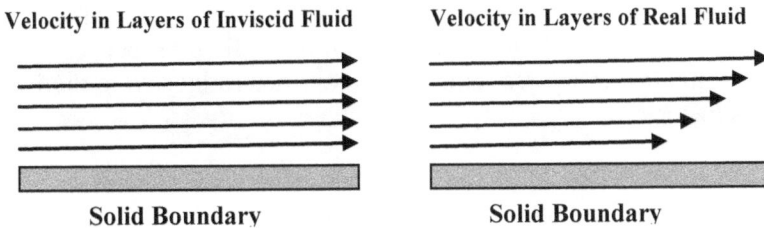

          Solid Boundary   **Solid Boundary**

The thickness of a boundary layer is defined, in practice, as that region of flowing fluid, starting from the fixed solid boundary, where the velocity or temperature of the fluid is 99 % of the average fluid velocity or temperature, respectively.

Heat transfer across the boundary layer is generally agreed to occur by conduction, so that if a boundary layer is of thickness, δ, the heat flux across it is $q = k \, \Delta T/\delta$, where k is the thermal conductivity of the fluid, and ΔT is the temperature difference across the boundary layer. δ is often difficult to measure directly so that a new parameter that is easier to measure, the film heat transfer coefficient, h, is defined such that $h = k/\delta$.

When the whole body of fluid is in laminar flow, the inertia forces, which move the fluid, are less than the viscous forces, which retard

15

the movement of fluid. In such cases, the fluid layers within the fluid move in an orderly and predictable manner so that the equations of fluid motion and heat transfer can be solved analytically.

The ratio of inertial forces to the viscous forces is called the Reynolds number, *Re*, and is the criterion which enables us to determine whether a fluid is flowing in laminar or turbulent flow. This ratio is defined as

$$\text{Re} = \frac{Inertia\,Forces}{Viscous\,Forces} = \frac{\rho U}{\dfrac{\mu}{D}} = \frac{\rho U D}{\mu} \tag{1.55}$$

U is the average or mean fluid velocity and D is a characteristic dimension of the conduit.

1.4.2: Turbulent Flow in Heat Convection

In turbulent flow, however, the inertia forces are several orders of magnitude greater than the viscous forces with the result that instability is introduced leading to fluctuating point velocities within the body of the fluid. Estimation of mean fluid properties now become statistical estimates rather than those determined from their previous history.

Whether the fluid is in laminar or turbulent flow, however, the common variables and objectives of interest are

      i). the velocity profile
      ii). the pressure drop
      iii). the shear stress at the solid boundary.

These variables enable us determine the volume, mass rate of flow or the required energy, to pump or withstand such a rate of flow in any fluid.

1.4.3: Internal Flow

Internal flow, such as flow through a pipe, occurs when fluid flow is enclosed by a solid boundary. Internal flow is important in forced convection. Velocity profiles for such flows are shown in Figures 1.3 and 1.4.

1.4.3.1: Laminar Internal Flow

The dotted lines trace the boundaries of the laminar layers from zero thickness at entry of fluid into the conduit to half the conduit diameter in fully developed laminar flow. The velocity profile, in laminar flow, is parabolic and is given by:

$$u = u_{\max}\left(1 - \frac{r^2}{R^2}\right) \qquad (1.56)$$

where u = fluid velocity at any radius, r; R = radius of the pipe and

u_{max} = the maximum velocity = $\dfrac{\Delta P\, R^2}{4\,\mu\, L}$

Figure 1.3: Velocity Profiles in Internal Laminar Flow

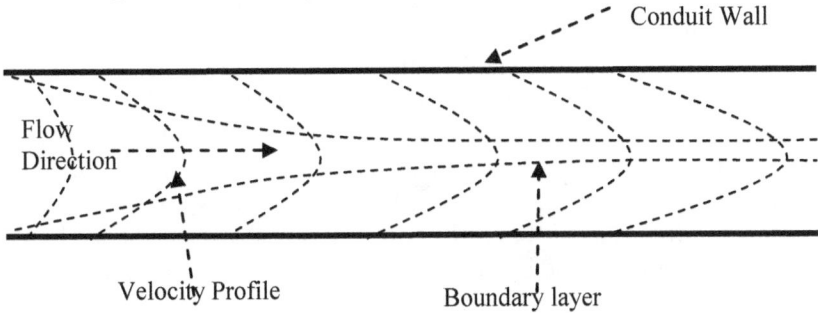

The average or mean velocity, u_{mean} is equal to one half of the maximum velocity, u_{max}, while the shear stress at the wall, τ_o is given by (Welty, 1978):

$$\tau_0 = \frac{8\,\mu\, u_{mean}}{D} \qquad (1.57)$$

The shear stress at the wall is, usually, expressed as a fraction of the free stream kinetic energy in order to obtain the friction coefficient. The friction factor is, generally, defined as

$$f = \frac{\tau_0}{\rho u_{mean}^2} \qquad (1.58)$$

The Fanning friction factor, C_f, is defined as

$$C_f = 2f = \frac{2\tau_0}{\rho u_{mean}^2}. \qquad (1.59)$$

That is:

$$\frac{C_f}{2} = f = \frac{\tau_0}{\rho u_{mean}^2} = \frac{16}{Re} \qquad (1.60)$$

The pressure drop, ΔP, is determined from the appropriate version of the Euler's equation (1.10) and is given by:

$$\Delta P = 2.C_f \cdot \frac{L}{D} \cdot \rho u_{mean}^2 \qquad (1.61)$$

C_f, the Fanning friction factor is, usually, plotted against the Reynolds number in the so called Moody chart (Welty, 1978). Charts for other cross-sections, configurations and metal matrices are also available (Kays and London, 1964).

1.4.3.2: Turbulent Internal Flow

In turbulent flow, the boundary layer thickness is a very small fraction of the radius of the pipe. The velocity, at any point in time, in the main body of the fluid, is the sum of some mean velocity and a fluctuating component. A time averaged mean velocity would, therefore, be expected to be constant even though the velocity, at any point and at any time, is not. This time averaged velocity, u, is the velocity used in turbulent flow calculations. The velocity profile is given by the Blasius seventh power law equation. For a circular conduit of radius R, the Blasius seventh power law equation is

$$\frac{u}{u_{max}} = \left(\frac{r}{R}\right)^{1/7} \qquad (1.62)$$

The shear stress at the wall is given, also by Blasius (Welty, 1978), as:

$$\frac{\tau_0}{\rho} = 0.0025 \frac{u_{max}^2}{\left(\frac{\rho u_{max} R}{\mu}\right)^{1/4}} \qquad (1.63)$$

where R is the radius of the conduit.

18

Figure 1.4: Velocity Profile in Internal Turbulent Flow

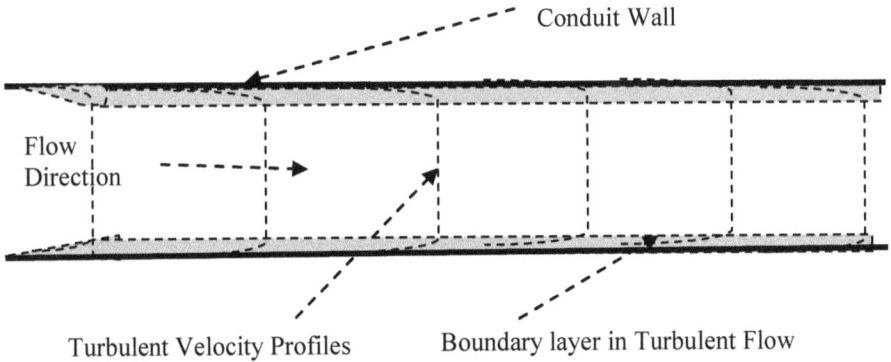

Turbulent Velocity Profiles   Boundary layer in Turbulent Flow

A common correlation of the friction factor with the Reynolds number, Re, pipe diameter, D, and surface roughness, ε, is given below:

$$\frac{1}{\sqrt{f}} = 2.457\ln\left(\frac{1}{\dfrac{0.888}{\mathrm{Re}\sqrt{f}} + 0.27\dfrac{\varepsilon}{D}}\right) \qquad (1.64)$$

In some countries, the Fanning friction factor is defined to be equal to the friction factor. That is $C_f = f$. This definition leads to the so called Colebrook equation given as

$$\frac{1}{\sqrt{f}} = -4\log\left(\frac{1.256}{\mathrm{Re}\sqrt{f}} + \frac{\varepsilon}{3.7D}\right) \qquad (1.65)$$

For smooth pipes, $\varepsilon = 0$, and the Colebrook equation reduces to the Prandtl equation given as

$$\frac{1}{\sqrt{f}} = -4\log\left(\frac{1.256}{\mathrm{Re}\sqrt{f}}\right) \qquad (1.66)$$

The pressure drop in turbulent flow is evaluated with the same equation for laminar flow (1.61) above.

1.4.3.3: Transition Flow in Internal Flow

The regime of flow between laminar and turbulent flow is known as the transition regime. Behaviour of fluid within this regime is more difficult to analyse since flow is, simultaneously, laminar and turbulent. The solution to this problem has been to set limits, based on experimental measurements and using the Reynolds number, to where laminar and turbulent flows begin or end.

For example, for flow inside circular pipes, below a Reynolds number of 2000, flow is laminar, above a Reynolds number of 10,000, flow is turbulent. Transition is expected to occur, therefore, between Reynolds numbers of 2000 and 10,000. That is

$$0 \leq Re \leq 2x10^3 \qquad \text{Laminar flow} \qquad (1.67)$$

$$2x10^3 \leq Re \leq 1x10^4 \qquad \text{Transition flow} \qquad (1.68)$$

$$Re \geq 1x10^4 \qquad \text{Turbulent flow} \qquad (1.69)$$

The need to develop one expression which is valid for all regimes led to the so called universal velocity profile. To obtain this profile, the space variable, y, and velocity variable, u, in the fluid flow equations, were put in dimensionless form as

$$y^* = \frac{y\sqrt{\frac{\tau_0}{\rho}}}{\nu} = \frac{y\sqrt{\rho\tau_0}}{\mu} \qquad (1.70)$$

$$u^* = \frac{\mu}{\sqrt{\frac{\tau_0}{\rho}}} \qquad (1.71)$$

The various regimes of flow, laminar, transition and turbulent, were, then, defined to occur when

$$0 < y^* \leq 5 \qquad u^* = y^* \qquad \text{Laminar Flow} \qquad (1.72)$$

$$5 < y^* \leq 30 \qquad u^* = 5\ln y^* - 3.05 \qquad \text{Transition Flow} \qquad (1.73)$$

$$y^* \geq 30 \qquad u^* = 2.5\ln y^* + 5.5 \qquad \text{Turbulent Flow} \qquad (1.74)$$

Recently, Churchill (1977) developed an expression for a friction factor applicable to all regimes which was defined as follows

$$f = \left[\left(\frac{8}{Re}\right)^{12} + \frac{1}{(A+B)^{3/2}} \right]^{1/12} \qquad (1.75)$$

where

$$A = \left[2.457\ln \left(\frac{1}{\left(\dfrac{7}{Re}\right)^{0.9} + 0.27\dfrac{\varepsilon}{D}} \right) \right]^{16}$$

(1.75a)

and

$$B = \left(\frac{37,530}{Re} \right)^{16}$$

(1.75b)

$\dfrac{\varepsilon}{D}$ is the relative roughness of the pipe.

1.4.4: External Flow

External flow occurs when fluid flows past a surface which does not have a closed circumference or perimeter. Such a surface could be a flat plate or an open channel. Laminar, turbulent and transition flows, also, take place as in internal flow except that the values of the transition Reynolds numbers, for the onset of each regime, are different from those for internal flow.

For example, for external flow parallel to a flat plate, the regimes are defined such that below a Reynolds number of 200, 000 (2 x 10^5), flow is laminar, above a Reynolds number of 3,000,000 (3 x 10^6), flow is turbulent. Transition flow occurs, therefore, between Reynolds numbers of 2 x 10^5 and 3 x 10^6. These may be summarised in tabular form as

$$0 \le Re_X \le 2x10^5 \qquad \text{Laminar flow} \qquad (1.76)$$

$$2x10^5 \le Re_X \le 3x10^6 \qquad \text{Transition flow} \qquad (1.77)$$

$$Re_X \ge 3x10^6 \qquad \text{Turbulent flow} \qquad (1.78)$$

The Reynolds number at any distance x, Re_X, is evaluated using the distance from the leading edge of the plate, the mean free stream velocity over the plate and the fluid physical properties.

For external flow past bluff or rounded objects, the regimes are determined, not only by the Reynolds number but also, by the angle subtended by the fluid flow on the surface of the object. For example, for flow perpendicular to a cylinder, the regimes are

21

specified as follows:

$0 \leq 85 \deg rees$	$\mathrm{Re}_D \leq 10^5$	Laminar flow	(1.79)
$85 \leq 0 \leq 135 \deg rees$	$\mathrm{Re}_D = 10^5$	Transition flow	(1.80)
$0 \geq 135 \deg rees$	$\mathrm{Re}_D \geq 10^5$	Turbulent flow	(1.81)

Re_D is the Reynolds number based on the cylinder diameter, D, and on the properties of the fluid. The friction coefficient, in this case, consists of the skin friction coefficient, due to viscous effects, C_f, and the form or pressure drag coefficient, due to pressure losses, C_D. C_D can only be obtained experimentally but both are shown as functions of the Reynolds number in Figures 1.5 and 1.6. The Moody (Fanning friction factor vs Re) chart for internal flow is shown in Figure 1.7.

1.5: Heat Transfer in Convection

The heat transferred in heat convection is often evaluated using one version or the other of the famous Newton's law of cooling. This law states that the rate of cooling of (or heat loss from) a hot surface, exposed in air, is directly proportional to the surface area and to the temperature difference between the surface and the air. This is, often, reformulated as

$$q = h(T_s - T_b)$$
(1.82)

q is the heat flux, h is a proportionality constant known as the heat transfer coefficient, T_s and T_b are the temperatures at the surface and bulk of fluid respectively. Recall that we had, earlier, defined a film heat transfer coefficient, in heat conduction, as k/δ.

Since $q = Q/A$ where Q is the heat transferred per unit time and A is the area through which heat is transferred, equation (1.82) can, also, be expressed as

$$Q = h A(T_s - T_b)$$
(1.82a)

Though developed for natural convection this formula has been found to be applicable also in forced convection provided the values of h, A and $(T_s - T_b)$ are evaluated correctly for the surface configuration encountered. The use of this value of h is valid because, in forced convection, a boundary layer is formed through which heat is still transferred by conduction. This tells us that, in all cases, h is always equal to k/δ.

h depends upon both the physical properties of the fluid (its thermal conductivity, *k*) and on the surface configuration in which convection occurs (which influences δ). Values of the heat transfer coefficient are, generally, obtained experimentally and presented as correlations for typical and, commercially, important configurations and fluids. Because of the wide variety of such configurations and fluids, discussion of heat transfer coefficients will be treated in the next chapter.

Fig. 1.5:   Drag Coefficients, C_D, (Cylinders) vs Reynolds Number in External Flow (Welty, 1978)

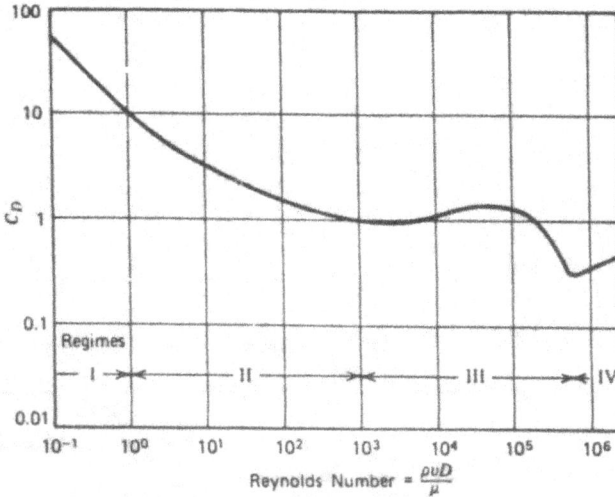

Fig. 1.6: Drag Coefficients, C_D, (Spheres and Other Shapes) vs Reynolds Number in External Flow (Welty, 1978)

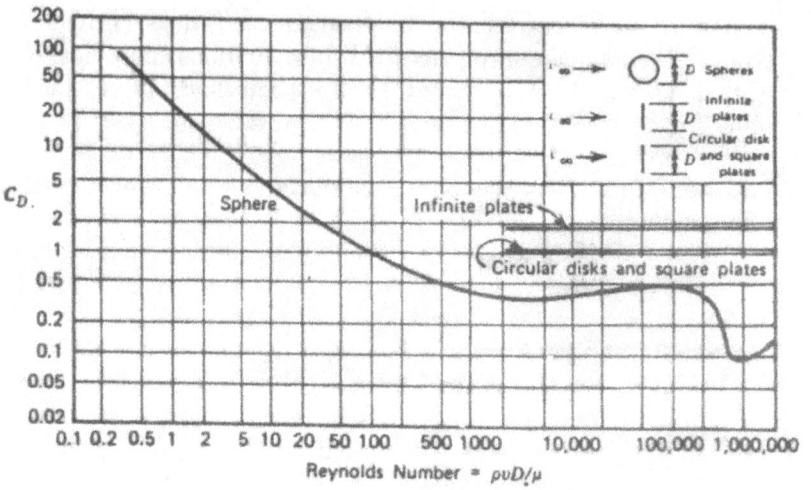

Fig. 1.7: Fanning Friction Factor, f, vs Re in Internal Flow (Welty, 1978)

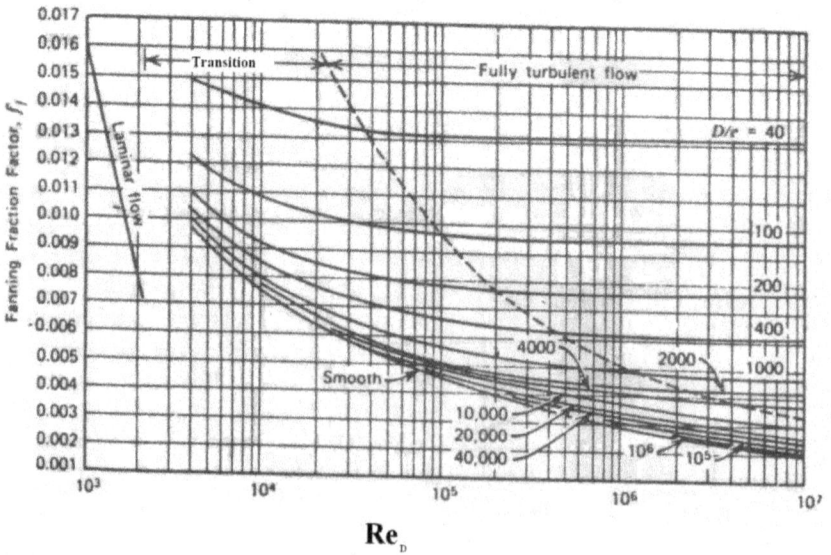

24

References for Chapter One

1   Churchill S. W., *Friction Factor Equation Spans all Fluid Flow Regimes*; Chem. Eng., Vol. 84, No.24; pp 91 - 92; McGraw-Hill Book Co., N. Y., U.S.A

2   Kay J. M; *An Introduction to Fluid Mechanics and Heat Transfer, 2nd. Edn.*, Cambridge University Press, London, 1965.

3   Kays W. M. and London A. L.; *Compact Heat Exchangers*, McGraw-Hill Book Company, New York, U.S.A., 1964.

4   Welty J. R.; *Engineering Heat Transfer, SI Version*, John Wiley and Sons, New York, U.S.A.,1978,

5   http://en.wikipedia.org/wiki/Derivation_of_the_Navier_Stokes_equa tions.

CHAPTER TWO
THE HEAT TRANSFER COEFFICIENT

2.1: The Heat Transfer Coefficient

We found, in the last chapter, that the heat, transferred in natural convection, could be estimated from

$$Q = h A(T_s - T_b) \qquad\qquad (1.82a)$$

as long as we can estimate, correctly, the values of **h**, A and $(T_s - T_b)$. Consider, however, a simple surface over which fluid is flowing, whether in natural or forced convection.

Figure 2.1 Heat Transfer Across and Along a Surface

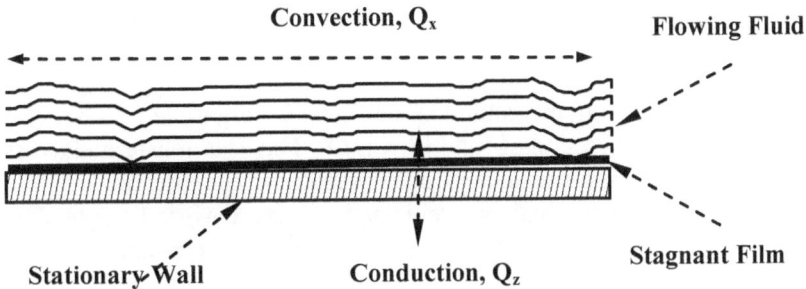

You may recall that, as this fluid flows past this surface, a boundary layer is formed and also a region outside the boundary layer in which bulk motion predominates. Within this region of bulk motion in the fluid, the dominant mode of heat transfer would be convection by means of eddies. A heat balance within this region and along the direction of fluid flow, *x*, gives

$$Q_x = m Cp \Delta T_x = \rho \, u \, A_x \, Cp \, \Delta T_x \qquad\qquad (2.1)$$

where Q_x, **m**, Cp are the heat transferred, the mass flow rate, in the x-direction, and heat capacity of the fluid, respectively, and **u**, A_x and ΔT_x are, respectively, the mean fluid velocity, conduit cross sectional area perpendicular to, and temperature difference in, the direction of flow.

In most cases of convective flow, the eddies die out in the vicinity of the wall thereby forming a thin stationary film of fluid which we know as the boundary layer. Heat transfer through this film is by conduction, because heat is transferred through this layer as if it was solid, and is perpendicular to the direction of fluid flow. If this layer is of thickness δ then by Fourier's first law of heat conduction

$$Q_z = k A_z \frac{\Delta T_z}{\delta} \tag{2.2}$$

where ΔT_z and A_z are, respectively, the temperature difference and area in the direction perpendicular to the direction of flow. Since the value of δ is generally not easily known, equation (2.2) is re-written as

$$Q_z = h A_z \, \Delta T_z \tag{2.3}$$

where $h = \dfrac{k}{\delta}$ is called the film heat transfer coefficient with units of W/m².K. Note the similarity of equation (2.3) to equation (1.82) which describes Newton's law of cooling

$$\frac{Q_z}{A_z} = q = h \, \Delta T_z \tag{2.3a}$$

The heat transfer coefficient is a very critical parameter in heat transfer analysis. Various methods of estimating or using it will be illustrated later in this chapter.

Reflect, however, on the situation depicted in Figure 2.1. The heat transferred in the direction of fluid flow is Q_x. The heat transferred across the surface of the solid boundary, perpendicular to the direction of fluid flow, is Q_z. If the energy balance depicted in Figure 1.1 is to be upheld, at steady state when no work is done and there is no accumulation of heat energy within the system,

$$Q_x = Q_z = \rho \, u \, A_x \, Cp \, \Delta T_x = h A_z \, \Delta T_z \tag{2.4}$$

This suggests the possibility of creating a heat transfer device by means of which we can transfer heat from one side of a surface, along which fluid is flowing, to the other side of the surface along which fluid may or may not be flowing. This device must ensure that the heat lost on one side of the surface is equal to that gained on the other side or vice versa. Such devices are known, generally, as heat exchangers.

Since all the other variables in equation (2.4) are easily determined, only the heat transfer coefficient poses a problem in determining the amount of heat energy transferred. Equation (2.4) also tells us that

$$h = \frac{\rho u \, A_x \, Cp \, \Delta T_x}{A_z \, \Delta T_z} \tag{2.4a}$$

Apart from ρ and Cp which are dependent only on the properties of the fluid, u, A_x, A_z and the two ΔT depend on the physical configuration of the surfaces involved in heat transfer.

Take, for example, the full flow of a fluid inside a cylindrical pipe of diameter, D. Over a length of the pipe, L, the ratio of A_x to A_z would be

$$\frac{A_x}{A_z} = \frac{\pi D^2}{4 \pi D L} = \frac{D}{4 L} \tag{2.5}$$

The heat transfer coefficient becomes

$$h = \frac{\rho u D Cp \Delta T_x}{4 L \Delta T_z} \tag{2.6}$$

For heat transfer through spherical surfaces of diameter, D, and thickness *x*, the equivalent expression is

$$h = \frac{\rho u D^2 Cp \Delta T_x}{(D+x)^2 \Delta T_z} \tag{2.7}$$

We could, probably estimate how the ΔTs are affected by surface configuration by means of our knowledge of the principles of heat conduction in solids. What is the value of *u* in equation (2.7)? What happens if another fluid is also flowing on the other side of the surface and at a different velocity? Could we determine these from fluid mechanics? How do we, generally, determine the heat transfer coefficient for the numerous situations for which we may need the transfer of heat?

Three approaches have found successful application in the estimation of the heat transfer coefficient. The first method estimates the heat transfer coefficient directly by experimental measurements. The second approach seeks to determine it by the mathematical solution of the original Navier-Stokes equations, with initial and boundary conditions defined by the situation at hand, and, then, confirming the result by experimental measurements. The third approach is to use dimensionless analysis, again, with the results confirmed by experimental measurements.

The continued success and improvements in these approaches were made possible, first, by the availability of mechanical computing machines, then the slide rule, then calculators and now digital computers. In modern times, simulation and computational fluid dynamics (CFD) make mathematical analysis almost routine for both regular and complex configurations.

The rest of this chapter will be devoted to the estimation and manipulation of the heat transfer coefficient in free and forced convection for both simple and complex surfaces using expressions derived from these methods.

2.1.1: The Overall Heat Transfer Coefficient

Consider a pipe of inner diameter, D_1, and outer diameter, D_2, through which a hot fluid is flowing in order to heat up a colder fluid flowing on the outside of it as shown in Figure 2.2. The mean temperature of the inside fluid is T_1, that of the outside fluid, T_4. The inner surface of the pipe is at the uniform temperature of T_2 while the outer surface of the pipe is at its uniform temperature of T_3.

Figure 2.2: Illustrative Diagram for Combination of Film Coefficients

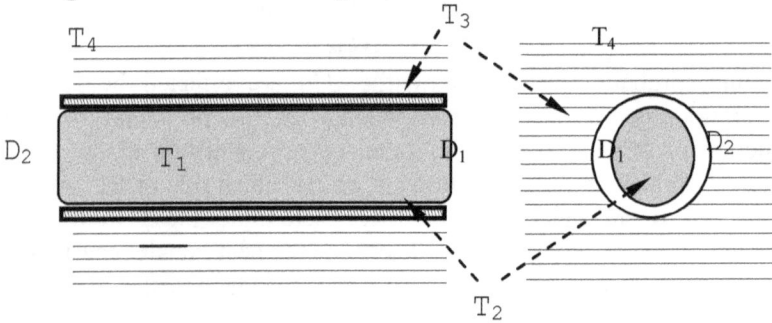

We shall designate the heat transfer coefficients, for the liquid film of fluid formed on each side of the tube, as h_1, and h_2, respectively. We must not forget that heat is transferred through the pipe wall by heat conduction. We calculate the heat transferred in the system as follows. Heat transferred inside the pipe along the direction of fluid flow is given, as before, by

$$Q_x = m Cp \Delta T_x = \rho \, u \, A_x \, Cp \, \Delta T_x \tag{2.1}$$

Heat transferred, from the hot fluid through the inner film layer to the inner surface of the pipe and in the direction of the radius of the pipe, is given, from equation (2.3) as

$$Q_z = h_1 A_1 \Delta T_{z1} = h_1 \pi D_1 L (T_1 - T_2) \tag{2.8}$$

Heat transferred, from the inner surface of the pipe through the thickness of the pipe to the outer surface of the pipe and still in the direction of the radius of the pipe, is given, from equation (2.2) as

$$Q_z = k A_z \frac{\Delta T_z}{z} = k A_m \frac{2(T_2 - T_3)}{D_2 - D_1} \tag{2.9}$$

where A_m is a mean surface area along the thickness of the pipe and z is the thickness of the wall of the pipe. If the pipe is very thin walled

$$A_m = \pi L \frac{(D_2 + D_1)}{2}, \text{ the arithmetic mean area} \tag{2.10}$$

Usually, however, the arithmetic mean area does not describe most cases accurately and the log mean surface area is used, given by

$$A_{lm} = \pi L \frac{(D_2 - D_1)}{\ln\left(\dfrac{D_2}{D_1}\right)}, \quad the \, log \, mean \, area \qquad (2.11)$$

For the situations where equation (2.10) can be used

$$Q_z = \pi k L \frac{D_2 + D_1}{D_2 - D_1} \cdot (T_2 - T_3) \qquad (2.12)$$

When equation (2.11) is used

$$Q_z = 2\pi k L \frac{(T_2 - T_3)}{\ln\left(\dfrac{D_2}{D_1}\right)} \qquad (2.13)$$

Heat transferred, from the outer surface of the pipe through the outer film layer and still in the direction of the radius of the pipe, is given, from equation (2.3), as

$$Q_z = h_2 A_2 \Delta T_{z2} = h_2 \pi D_2 L (T_3 - T_4) \qquad (2.14)$$

From equation (2.8)

$$T_1 - T_2 = \frac{Q_z}{h_1 A_1} = \frac{Q_z}{h_1 \pi D_1 L} \qquad (2.8a)$$

From equation (2.9) and (2.13)

$$T_2 - T_3 = \frac{D_2 - D_1}{2k A_{lm}} = \frac{Q_z}{2\pi k L} \ln\left(\frac{D_2}{D_1}\right) \qquad (2.13a)$$

From equation (2.14)

$$T_3 - T_4 = \frac{Q_z}{h_2 A_2} = \frac{Q_z}{h_2 \pi D_2 L} \qquad (2.14a)$$

Adding equations (2.8a), (2.13a) and (2.14a)

$$T_1 - T_4 = Q_z \left[\frac{1}{A_1 h_1} + \frac{D_2 - D_1}{2k A_{lm}} + \frac{1}{A_2 h_2} \right] \qquad (2.15)$$

or

$$T_1 - T_4 = \frac{Q_z}{\pi D_1 L} \left[\frac{1}{h_1} + \frac{D_1}{2k} \ln\left(\frac{D_2}{D_1}\right) + \frac{D_1}{h_2 D_2} \right] \qquad (2.15a)$$

If we define

$$Q_z = U_1 A_1 \Delta T = U_2 A_2 \Delta T \qquad (2.16)$$

where $\Delta T = T_1 - T_4$ and U_1 is an overall heat transfer coefficient based on surface area 1 and U_2 is an overall heat transfer coefficient based on

31

surface area 2, we can see that

$$\frac{\Delta T}{Q_z} = \frac{1}{U_1 A_1} = \frac{1}{U_2 A_2} = \left[\frac{1}{A_1 h_1} + \frac{D_2 - D_1}{2k A_{lm}} + \frac{1}{A_2 h_2}\right] \qquad (2.17)$$

This tells us that

$$\frac{1}{U_1} = \frac{A_1}{U_2 A_2} = \left[\frac{1}{h_1} + \frac{D_2 - D_1}{2k}\cdot\frac{A_1}{A_{lm}} + \frac{A_1}{A_2 h_2}\right] \qquad (2.18)$$

and that

$$\frac{1}{U_2} = \frac{A_2}{U_1 A_1} = \left[\frac{A_2}{A_1 h_1} + \frac{D_2 - D_1}{2k}\cdot\frac{A_2}{A_{lm}} + \frac{1}{h_2}\right] \qquad (2.19)$$

These may be expressed in terms of pipe diameters as

$$\frac{1}{U_1} = \frac{D_1}{U_2 D_2} = \left[\frac{1}{h_1} + \frac{D_1}{2k}\ln\left(\frac{D_2}{D_1}\right) + \frac{D_1}{h_2 D_2}\right] \qquad (2.20)$$

and as

$$\frac{1}{U_2} = \frac{D_2}{U_1 D_1} = \left[\frac{D_2}{h_1 D_1} + \frac{D_2}{2k}\ln\left(\frac{D_2}{D_1}\right) + \frac{1}{h_2}\right] \qquad (2.21)$$

Illustrative Example 2.1

A composite wall is made up of an external thickness of brickwork 110mm thick inside which is a layer of fiber glass 75 mm thick. The fiber glass is faced internally by an insulating board 25 mm thick. The coefficients of thermal conductivity for the three materials are:

Brickwork	1.15 W/mK
Fiber glass	0.04 W/mK
Insulating Board	0.06 W/mK

The inside and outside film surface transfer coefficients are 2.5 W/mK and 3.1 W/mK respectively. Determine the overall heat transfer coefficient and heat loss through such a wall 6m high and 10m long. The internal and external ambient temperatures are $27°$ C and $10°$ C respectively.

Answer

A section of the wall may be represented as shown below. Let the insulating board be labelled by 1, fiberglass 2 and building brick 3. Since the walls are parallel,

32

$$A_1 = A_2 = A_3 = A_m = 6 \times 10 = 60 \text{ m}^2.$$

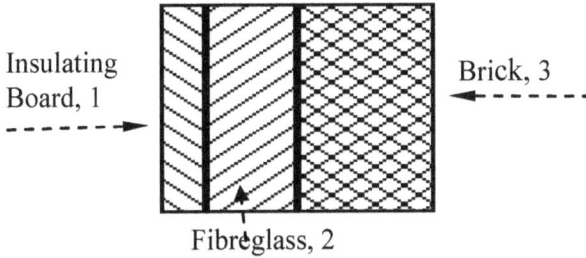

Insulating Board, 1

Brick, 3

Fibreglass, 2

$$\Delta T = (27 + 273) - (10 + 273) = 17 \text{ K}.$$

$$\frac{1}{U_1} = \frac{1}{h_1} + \frac{x_1}{k_1} + \frac{x_2}{k_2} + \frac{x_3}{k_3} + \frac{1}{h_2}, \frac{m^2 K}{W}$$

$$= \frac{1}{2.5} + \frac{0.025}{0.06} + \frac{0.075}{0.04} + \frac{0.11}{1.15} + \frac{1}{3.1}, \frac{m^2 K}{W}$$

$$= 0.4 + 0.417 + 1.875 + 0.096 + 0.323 \frac{m^2 K}{W} = 3.111 \frac{m^2 K}{W}$$

That is $U_1 = 0.321, \dfrac{W}{m^2 K}.$ and

$$Q = U_1 A_m \Delta T = 0.321 \frac{W}{m^2 K} \times 60 \, m^2 \times 17 K = 327.42 W \; Ans$$

2.2: Evaluating Film Heat Transfer Coefficients

Film heat transfer coefficients are, in practice, generally, estimated, from measurements, dimensionless expressions, and dimensional expressions, all based on, or confirmed by, experimental results

Numerical data, presented in tables, often apply to stated specific situations and unless the situation you are dealing with is similar, are useful only as guides. A few typical numerical values are given in Table 2.1.

The dimensionless expressions are much more general in application and are, essentially, based on correlations between two or three of several dimensionless groups listed in Tables 2.2 and 2.3.

33

Table 2.1: Typical Numerical Values of Film Heat Transfer Coefficients (Welty, Wilson & Wicks, 1976)

System	Film Heat Transfer Coefficient, h, W/m^2K
Free Convection, Air	5 – 50
Forced Convection, Air	25 – 250
Forced Convection, Water	250 – 15,000
Boiling, Water	2,500 – 25,000
Condensing, Water	500 – 100,000

Table 2.2: Summary of the Main Dimensionless Groups in Fluid Mechanics (Perry & Green, 1984)

Name	Symbol	Formula	Special Notes
Momentum Diffusivity	ν	$\dfrac{\mu}{\rho}$	Also known as Stoke's viscosity
Bingham Number	B_m or N_{Bm}	$\dfrac{\tau_y L}{\mu_p u}$	τ_y = yield stress μ_p = coefficient of rigidity
Cauchy Number	C or N_C	$\dfrac{\rho u^2}{E_b}$	E_b = Bulk Modulus
Drag Coefficient	C_D	$\dfrac{(\rho_p - \rho_f)g L}{\rho_p u^2}$	ρ_p = density of object ρ_f = density of fluid
Euler Number	Eu or N_{Eu}	$\dfrac{\Delta P_f}{\rho u^2}$	$\dfrac{\Delta P_f}{\rho}$ = friction heat
Fanning Friction Factor	f or C_f	$\dfrac{\Delta P_f . D}{2 u^2 L}$	D = pipe diameter L = length of pipe
Froude Number	Fr or N_{Fr}	$\dfrac{u^2}{g L}$	
Mach Number	Ma or N_{Ma}	$\dfrac{u}{u_C}$	u_C = velocity of sound in fluid
Reynolds Number	Re or N_{Re}	$\dfrac{\rho D u}{\mu}$	
Weber Number	We or N_{We}	$\dfrac{\rho L u^2}{\sigma}$	σ = surface tension

Table 2.3: Summary of the Main Dimensionless Groups in Heat Transfer (Perry & Green, 1984)

Name	Symbol	Formula
Thermal Diffusivity	α	$\dfrac{k}{\rho Cp}$
Biot Number	Bi or N_{Bi}	$\dfrac{hL}{k}$
Fourier Number	Fo or N_{Fo}	$\dfrac{kt}{\rho Cp L^2}$
Graetz Number	Gz or N_{Gz}	$Gr = Re.Pr.\dfrac{D}{L}$
Grashof Number	Gr or N_{Gr}	$\dfrac{L^3 \rho^2 g\beta \Delta T}{\mu^2}$
Nusselt Number	Nu or N_{Nu}	$\dfrac{hD}{k}$
Peclet Number	Pe or N_{Pe}	$\dfrac{\rho u Cp L}{k}$
Prandtl Number	Pr or N_{Pr}	$\dfrac{Cp\,\mu}{k}$
Stanton Number	St or N_{St}	$\dfrac{h}{\rho u Cp}$
Rayleigh Number	Ra or N_{Ra}	$Ra = Gr.Pr = \dfrac{g\beta(T_s - T_\infty)L^3}{\nu\alpha}$

In addition to the regimes of flow encountered in fluid flow such as laminar, transition and turbulent flow, heat transfer in fluids has its own regimes, the first of which we saw as free and forced convection. A further classification is in terms of whether the heat transfer takes place with or without phase change. This latter classification is particularly important in estimating the heat transfer coefficient when there is a change in phase as the heat flux behaviour, as a function of temperature difference is, remarkably, different from its behaviour when there is no phase change.

Because temperature difference levels in free convection are not high enough to cause significant phase change, it is, usually, assumed that phase change is important only in forced convection. Nevertheless, we shall start the estimation of heat transfer coefficients first for free convection, then for forced convection without phase change and then for forced convection with phase change.

2.2.1: Heat Transfer Coefficients in Free Convection

Dimensionless analysis, confirmed by experiment, shows relationships between the Nusselt, Grashof and Prandtl numbers especially for heat transfer in air and in certain, common, physical configurations. This relationship is of the form $Nu = f(Gr.Pr)$. Thus, if the Nusselt number can be evaluated from these relationships, the film heat transfer coefficient can then be estimated from

$$h = Nu . \frac{k}{D} \qquad (2.2)$$

We shall, now, consider some of these relationships between the Prandtl and Grashof numbers and the configurations for which they have been determined.

2.2.1.1: Vertical Heated Plates And Cylinders In Natural Convection

Eckert and Jackson are reported in Welty (1978) to have found that for plates and cylinders of characteristic length, L,

$$Gr.Pr<10^9, \quad Nu_L = 0.555(Gr.Pr)^{\frac{1}{4}} \qquad (2.3)$$

$$Gr.Pr > 10^9, \quad Nu_L = 0.021(Gr.Pr)^{\frac{2}{5}} \qquad (2.4)$$

For constant wall temperature,

$$Nu_L = 0.678 \frac{Pr^{\frac{1}{2}} Gr_L^{\frac{1}{4}}}{(0.952 + Pr)^{\frac{1}{4}}} \qquad (2.5)$$

2.2.1.2: Natural Convection in Vertical Channels

For vertical channels with constant wall temperatures and in which one surface is heated and the other cooled, the relationship is found to include the aspect ratio of the channel in addition to the Nusselt, Grashof and Prandtl numbers. The aspect ratio is simply the ratio of the height, H, to the width, L, of the channel. That is $Nu = f(Gr, Pr, H/L)$.

36

According to Welty (1978) Jakob correlated the experimental data of other workers and reported that three regions are observed in natural convection in this system. These regions and their characteristics are listed in Table 2.4.

2.2.1.3: Natural Convection on Horizontal Curved Surfaces

Mc Adams (1954) reported that, for heated horizontal solid cylinders of diameter D, in liquids and gases, when $10^4 < Gr_D.Pr < 10^9$,

$$Nu_D = 0.53(Gr_D.Pr)^{\frac{1}{4}} \qquad (2.9)$$

When $Gr_D.Pr < 10^4$ and the cylinder diameter is very small, such as in a wire, Elenbaas (Welty, 1978) gave the expression

$$Nu_D^3.e^{\frac{6}{Nu}} = \frac{Gr_D.Pr}{235} \qquad (2.10)$$

For $Gr_D.Pr < 10^4$ but the cylinder diameter larger than that of a wire, Hsu (Welty, 1978) gave the expression

$$Nu_D = 0.53\frac{(Gr_D.Pr)^{\frac{1}{4}}}{(0.952 + Pr)^{\frac{1}{4}}} \qquad (2.11)$$

Table 2.4: Values of the Nusselt Number for Natural Convection in Vertical Channels

Grashof Number Range	Dominant Mode of Fluid and Heat Flow	Value of Nusselt Number
$Gr_L \le 2 \times 10^3$	Conduction	$Nu_L = 1.0 \quad i.e$ $\quad h = \frac{k}{L} \qquad (2.6)$
$2 \times 10^3 < Gr_L \le 2 \times 10^5$	Laminar Flow	$Nu_L = 0.18 Gr^{\frac{1}{4}}\left(\frac{H}{L}\right)^{-\frac{1}{9}} \quad (2.7)$
$2 \times 10^5 < Gr_L \le 2 \times 10^7$	Turbulent Flow	$Nu_L = 0.065 Gr^{\frac{1}{3}}\left(\frac{H}{L}\right)^{-\frac{1}{9}} \quad (2.8)$

2.2.1.4: Natural Convection On Horizontal Plane Surfaces

Mc Adams (Welty, 1978) gives for
   (a) Hot plates facing up or cold plates facing down, in the range
   $10^5 < Gr_L . Pr < 2 \ x \ 10^7$,

$$Nu_L = 0.54(Gr_L . Pr)^{\frac{1}{4}} \qquad (2.12)$$

and in the range $2 \ x \ 10^7 < Gr_L . Pr < 3 \ x \ 10^{10}$,

$$Nu_L = 0.14(Gr_L . Pr)^{\frac{1}{2}} \qquad (2.13)$$

   (b) Hot plates facing down and cold plates facing up, in the range
   $3 \ x \ 10^5 < Gr_L . Pr < 1 \ x \ 10^{10}$,

$$Nu_L = 0.27(Gr_L . Pr)^{\frac{1}{4}} \qquad (2.14)$$

The characteristic length here, L, is, depending on the surface configuration, equal to either the length of a side of a square surface or the mean of the dimensions of a rectangular surface or 0.9 times the diameter of a circular area.

2.2.1.5: Natural Convection in Spheres and Rectangular Solids

For spheres and rectangular solids, the expressions for horizontal cylinders are used with a modified characteristic length given as

$$\frac{1}{L} = \frac{1}{L_{horizontal}} + \frac{1}{L_{vertical}} \qquad (2.15)$$

For a sphere, for example

$$\frac{1}{L} = \frac{1}{L_{horizontal}} + \frac{1}{L_{vertical}} = \frac{1}{D} + \frac{1}{D} = \frac{2}{D} \quad or \quad L = \frac{D}{2} \quad (2.15a)$$

2.2.1.6: Natural (Free) Convection in any Geometry

A popular correlation for determining the heat transfer coefficient, and hence the heat transferred, in natural convection and which applies to a variety   of   geometries   is

38

$$Nu - \left[Nu_o^{1/2} + Ra^{\frac{1}{6}} \left(\frac{f_4 \cdot Pr}{300} \right)^{\frac{1}{6}} \right]^2 \tag{2.16}$$

Nu is the Nusselt number, **Ra** is the Rayleigh number, and **Pr** is the Prandtl number. The value of $f_4 \cdot Pr$ is calculated using the following formula

$$f_4 \cdot Pr = \left[1 + \left(\frac{0.5}{Pr} \right)^{\frac{9}{16}} \right]^{-\frac{16}{9}} \tag{2.17}$$

The values of **Nu₀** and the characteristic length used to calculate **Ra** are listed below:

Geometry	Characteristic Length	Nu₀
Inclined Plane	Distance along plane	0.68
Inclined Disk	$\dfrac{9D}{11}$ (D = Diameter)	0.56
Vertical Cylinder	height of cylinder	0.68
Cone	$\dfrac{4X}{5}$ (X = distance along sloping surface)	0.54
Horizontal Cylinder	$\dfrac{\pi D}{2}$ (D = Diameter of cylinder)	0.36π

2.2.1.7: Natural Convection in Air at Atmospheric Pressure

The majority of actual cases of natural convection involve air. Since the general form of the expressions is

$$Nu_L = a^* (Gr_L \cdot Pr)^{b^*} \tag{2.18}$$

where a^* and b^* are constants, it is easily seen, that on expressing the dimensionless groups in full,

$$\frac{hL}{k} = a^* \left(\frac{L^3 \rho^2 \, g\beta \, \Delta T}{\mu^2} \cdot \frac{Cp\,\mu}{k} \right)^{b^*} \tag{2.18a}$$

only L and T do not involve the properties of air. Equation (2.18a) can, therefore, be reduced to a more convenient form

$$h = a\left(\frac{\Delta T}{L}\right)^b \qquad (2.19)$$

in which **a** (incorporating all the fluid properties) and **b** are still constants and $\Delta T = T_{surface} - T_{bulk\ air}$. The constants, **a** and **b,** as evaluated by Mc Adams (Welty, 1978), are summarized in Table 2.5.

Table 2.5: Values of Constants to be used in $h = a\left(\frac{\Delta T}{L}\right)^b$

Geometry	Applicable Range	a	b	L
Vertical surfaces (planes & cylinders)	$10^9 < Gr_L . Pr < 10^{12}$	1.42 1.31	1/4 1/3	= height = 1
Horizontal cylinders	$10^3 < Gr_L . Pr < 10^9$ $10^9 < Gr_L . Pr < 10^{12}$	1.32 1.24	1/4 1/3	= diameter = 1
Horizontal planes: hot plate facing up	$10^5 < Gr_L . Pr < 2 \times 10^7$	1.32	1/4	= length of side
Horizontal planes: Cold plate facing down	$2 \times 10^7 < Gr_L . Pr < 3 \times 10^{10}$	1.52	1/3	= 1
Horizontal planes: Cold plate facing up or hot plate facing down	$3 \times 10^5 < Gr_L . Pr < 3 \times 10^{10}$	0.59	1/4	= length of side

2.2.2: Heat Transfer Coefficients in Forced Convection, No Phase Change

As in free convection but much more so, the film heat transfer coefficient, in forced convection, is affected by whether the fluid flow is internal or external, laminar, transition or turbulent.

Internal Flow

Most cases of heat transfer in internal flow, which are of commercial importance, are connected with flow inside circular tubes and pipes. Two limiting temperature conditions, commonly encountered, and between which many variations can occur, are those in which

40

a. the pipe wall is at a constant temperature such as the temperature of condensing steam
b. there is a constant heat flux across the pipe wall such as occurs when an electric heating element provides the heat energy

For these two conditions and their many variants, film heat transfer coefficients are evaluated using the following expressions.

2.2.2.1: Laminar Internal Flow Heat Transfer at Constant Wall Temperature

For fully developed laminar flow, analytical solution of the Navier-Stokes equations gives

$$Nu = \frac{hD}{k} = 3.658 \tag{2.20}$$

Seider and Tate (Welty, 1978) obtained by experiment

$$Nu_L = 1.86 \left(\frac{Re.Pr.D}{L} \right)^{\frac{1}{3}} \left(\frac{\mu_m}{\mu_s} \right)^{0.14} = 1.86 Grz^{\frac{1}{3}} \left(\frac{\mu_m}{\mu_s} \right)^{0.14} \tag{2.21}$$

To use equation (2.21), all fluid properties are evaluated at the average of the mean fluid temperature except μ_s which is evaluated at the surface temperature T_s. Churchill (1977) developed an expression which is claimed to be valid throughout the whole of the laminar flow regime. This is

$$Nu = 3.657 \left[1 + \left(\frac{Grz}{7.6} \right)^{\frac{8}{3}} \right]^{\frac{1}{8}} \tag{2.22}$$

2.2.2.2: Laminar Internal Flow Heat Transfer with Constant Wall Heat Flux

Here, $q_{surface} = h(T_{surface} - T_{mean}) = constant$. When $Grz < 100$, the Hansen equation is used (Perry & Green, 1984)

$$Nu_{lm} = 4.364 + \frac{0.085 Grz}{1 + 0.047 Grz^{\frac{2}{3}}} \left(\frac{\mu_m}{\mu_s} \right)^{0.14} \tag{2.23}$$

Nu_{ulm} is the log mean of the Nusselt numbers at the two conditions of interest. The Graetz number and μ_m are evaluated at the mean fluid temperature while μ_s is evaluated at the mean wall temperature.

41

When Grz > 100, an analytical solution of the Navier-Stokes equations gives

$$Nu_D = \frac{hD}{k} = 4.364 \tag{2.24}$$

For small diameters, small temperature differences and Grz > 100, the Seider and Tate equation (2.21) above is used but with an arithmetic mean value of the Nusselt number, Nu_{am}. That is

$$Nu_{am} = 1.86 Grz^{\frac{1}{3}} \left(\frac{\mu_m}{\mu_s}\right)^{0.14} \tag{2.25}$$

For all diameters and temperature differences and still for Grz > 100, the recommended equation is

$$Nu_{am} = 1.86 Grz^{\frac{1}{3}} \left(\frac{\mu_m}{\mu_s}\right)^{0.14} + 0.87\left(1 + 0.015 Grz^{\frac{1}{3}}\right) \tag{2.26}$$

An expression, developed by Churchill (1977), which is claimed to be applicable throughout the laminar flow regime (Re < 2100), is

$$Nu = 4.364\left[1 + \left(\frac{Grz}{7.3}\right)^2\right]^{\frac{1}{6}} \tag{2.27}$$

2.2.2.3: Turbulent Internal Flow Heat Transfer with Constant Wall Temperature

No simple analytical solutions exist on account of the complex nature of turbulence. Empirical correlations are, therefore, commonly used. Depending on the English speaking area, these correlations are given, sometimes with different or slightly different names.

i. the Dittus - Boelter equation (known, also, as the Nusselt equation) is given as

$$Nu_D = 0.023 Re_D^{0.8} Pr^n \tag{2.28}$$

where $n = 0.3$ for fluid being cooled,
$n = 0.4$ for fluid being heated.

Equation (2.28) is valid in the range Re > 10^4 and in the range of Prandtl numbers $0.7 < Pr < 100$.

ii. the Colburn equation

$$St = \frac{h}{\rho Cp u} = 0.023 Re_D^{-0.2} Pr^{-\frac{2}{3}} \tag{2.29}$$

The Stanton number, St, is evaluated at the average of the mean bulk temperatures of the fluid while the Reynolds (Re) and Prandtl (Pr) numbers are evaluated at the average fluid film temperatures. The Colburn equation applies when Re $> 10^4$, in the range of Prandtl numbers $0.7 < \text{Pr} < 160$ and aspect ratio L/D > 60. The Colburn equation is, also, known as the Chilton-Colburn equation. Sometimes, equation (2.29) is rearranged in the form

$$j_h = 0.023\text{Re}^{-0.2} = St.\text{Pr}^{\frac{2}{3}} \qquad (2.29a)$$

as the so called j-factors. The j_h is the j-factor for heat transfer. There is a j_D factor for mass transfer. The idea is to model turbulent heat transfer after turbulent momentum transfer and to get variables similar to the Fanning or other friction factors.

iii. McAdam's equation

This is a modification of the Colburn equation using the Seider and Tate type viscosity correction factor. The modified equation is

$$St = 0.023\text{Re}_D^{-0.2}\,\text{Pr}^{-\frac{2}{3}}\left(\frac{\mu_m}{\mu_s}\right)^{0.14} \qquad (2.30)$$

All fluid properties are evaluated at the average fluid mean temperatures except μ_s which is evaluated at the wall temperature. Equation (2.30) applies when Re $> 10^4$, $0.7 < \text{Pr} < 17{,}000$ and aspect ratio L/D > 60.

iv. Coulson, Richardson & Sinnott, (1983) state that the Chilton – Colburn, Dittus – Boelter, McAdam or Seider Tate equations are, in reality, different forms of the general expression

$$Nu = C\,\text{Re}^{0.8}\,\text{Pr}^{0.33}\left(\frac{\mu_m}{\mu_s}\right)^{0.14} \qquad (2.31)$$

where C = 0.021 for gases
= 0.023 for non-viscous liquids
= 0.027 for viscous liquids

v. the Engineering Sciences Data Unit (ESDU) expression is

$$St = E.\text{Re}^{-0.205}\,\text{Pr}^{0.505} \qquad (2.32)$$

where

$$E = 0.0225\exp\left(-0.0225(\ln \text{Pr})^2\right) \qquad (2.32a)$$

43

This expression is claimed by Butterworth, (1977), as reported in Coulson, Richardson & Sinnott, (1983), to give more accurate estimates of heat transfer coefficients than the previous expressions.

vi. Churchill (1977) developed expressions which is claimed to be applicable to all regimes of fluid flow in smooth, circular pipes, provided flow is fully developed and, approximately, isothermal. These expressions are

$$
Nu^{10} = Nu_L^{10} + \left[\frac{\exp\left(\dfrac{2200 - Re}{365}\right)}{Nu_{LC}^2} + \left(\frac{1}{Nu_0^0 + \dfrac{0.079 Re\sqrt{f}.Pr}{\left(1 + Pr^{\frac{4}{5}}\right)^{\frac{5}{6}}}} \right)^2 \right]^{-5}
\tag{2.33}
$$

where Nu_{LC} is the critical Nusselt number and

$$
Nu_L = 3.657 \left[1 + \left(\frac{Grz}{7.60} \right)^{\frac{8}{3}} \right]^{\frac{1}{8}}
\tag{2.34}
$$

As Re → 2100, Pr → 0 and Nu → Nu_∞, Nu_∞ = 5.76 (from theory) and Nu_∞ = 4.8 (from experiments).

For Re ≥ 2100, Nu_{LC}, the Nusselt number at the transition from laminar to turbulent flow, is given by

$$
Nu_{LC} = 3.657 \left[1 + \left(\frac{276 Pr.D}{L} \right)^{\frac{8}{3}} \right]^{\frac{1}{8}}
\tag{2.35}
$$

The friction factor, f, in isothermal flow, is evaluated for each regime using either

i   Re < 2100 $f = \dfrac{8}{Re}$ (2.36)

ii $2200 < \text{Re} < 2700$ $f = \left(\dfrac{\text{Re}}{36{,}500}\right)^2$ (2.37)

iii $\text{Re} > 2700$ $\dfrac{1}{f} = 2.21\ln\left(\dfrac{\text{Re}}{7}\right)$ (2.38)

or, for all regimes of flow, (Churchill, 1977),

$$\frac{1}{f} = \left[\frac{1}{\left\{\left(\dfrac{8}{\text{Re}}\right)^{10} + \left(\dfrac{\text{Re}}{36{,}500}\right)^{20}\right\}^{\frac{1}{2}}} + \left\{2.21\ln\left(\dfrac{\text{Re}}{7}\right)\right\}^{10}\right]^{\frac{1}{5}}$$ (2.39)

2.2.2.4: Heat Transfer with Constant Wall Heat Flux in Turbulent Internal Flow

For constant wall heat flux, Churchill's equation (2.33) still applies but with Nu_{LC}, the critical Nusselt number, given by

$$Nu_L = 4.364\left[1 + \left(\frac{Grz}{7.30}\right)^2\right]^{\frac{1}{6}}$$ (2.40)

As $\text{Re} \to 2100$, $\text{Pr} \to 0$, and $\text{Nu} \to \text{Nu}_\infty$, $\text{Nu}_\infty = 8$ (from theory) and $\text{Nu}_\infty = 6.3$ (from experiments).

For $\text{Re} \geq 2100$, Nu_{LC}, the Nusselt number at the transition from laminar to turbulent flow, is given by

$$Nu_{LC} = 4.364\left[1 + \left(\frac{287\text{Pr}.D}{L}\right)^2\right]^{\frac{1}{6}}$$ (2.41)

2.2.2.5: Turbulent Internal Flow Heat Transfer in Non-Isothermal Flow

In non-isothermal flow, Churchill's equation (2.33) still applies but Nu_L and Nu_{LC}, are based on fluid properties evaluated at the mixed fluid mean temperature. The Prandtl number is evaluated at the wall surface temperature. The Reynolds number is evaluated at the average of the surface and mixed fluid temperatures except for the Reynolds number in the exponential term which is evaluated at the mixed fluid mean temperature.

In laminar but non-isothermal flow, the Nusselt number needs to be corrected by the ratio of the viscosity of the fluid, evaluated at the surface temperature, to that evaluated at the mixed fluid mean temperature.

The friction factor, f, in non-isothermal flow, is evaluated as for isothermal flow except that the Reynolds number is evaluated based on the surface temperature.

External Flow

In evaluating film heat transfer coefficients in external flow, the approach used, except in very simple configurations, is to obtain correlations which are similar to those used in internal flow. This approach is, especially, useful in the design and analysis of heat exchangers, to be discussed later.

The delineation of the regimes of laminar or turbulent flow are, however, not as straightforward as in internal flow, except in very simple configurations. Constant wall surface temperature and constant wall heat flux as well as non-isothermal conditions still occur.

A few of the more typical and frequently employed flow and surface configurations are discussed below. These configurations are those which appear most likely to be useful, or in common use, in convective heat transfer between one fluid or system and another. Often these involve plane, cylindrical and spherical surfaces in the form of flat plates and walls, tubes, pipes and spherical objects.

2.2.3.1: Heat Transfer about Plane Surfaces in External Flow

When the fluid is in laminar flow over a flat isothermal plate, the film heat transfer coefficient may be evaluated using

$$Nu_L = 0.664. \Pr^{\frac{1}{3}} . \sqrt{\operatorname{Re}_L} \qquad (2.42)$$

Mean film temperatures are used in evaluating fluid properties. When the flow is turbulent

$$Nu_L = 0.036. \Pr^{\frac{1}{3}} . \operatorname{Re}_L^{\frac{4}{5}} \qquad (2.43)$$

In transition flow, a combination of the two equations is used. These

46

equations are valid in the Prandtl number range; $0.6 < \text{Pr} < 50$

2.2.3.2: Heat Transfer about Spheres in External Flow

For this, the empirical correlations of McAdams are used. These are summarized in Table 2.6. Fluid properties are evaluated at the mean film temperature.

Table 2.6: Film Heat Transfer Correlations for External Flow about Spheres

Fluid	Reynolds Number Range	Correlation	
Liquids	$1 < \text{Re}_D < 70{,}000$	$Nu_D = 2.0 + 0.60 . \text{Pr}^{\frac{1}{3}} . \sqrt{\text{Re}_D}$	(2.44)
Air	$20 < \text{Re}_D < 150{,}000$	$Nu_D = 0.33 . \text{Re}_D^{0.6}$	(2.45)
Gases other than Air	$1 < \text{Re}_D < 25$	$St = \dfrac{2}{\text{Re}_D} + \dfrac{0.48}{\sqrt{\text{Re}_D}}$	(2.46)
	$25 < \text{Re}_D < 150{,}000$	$Nu_D = 0.37 . \text{Pr}^{\frac{1}{3}} . \text{Re}^{0.6}$	(2.47)

2.2.3.3: Heat Transfer across Single Cylinders in External Flow

For flow across single cylinders, McAdams (Welty, 1978) gives,

for gases $$Nu_D = B . \text{Re}_D^n \tag{2.48}$$

and for liquids $$Nu_D = 1.1 . B . \text{Re}_D^n . \text{Pr}^{\frac{1}{3}} \tag{2.49}$$

where B and n are constants given below(Table 2.7)

Table 2.7: Value of Constants for the McAdam Equation

Re_D	B	n
0.4 - 4	0.891	0.330
4 - 40	0.821	0.385
40 - 4,000	0.615	0.466
4,000 - 40,000	0.174	0.618
40,000 - 400,000	0.0239	0.805

2.2.3.4: External Flow Heat Transfer across Banks of Tubes

Many convective heat exchange devices employ banks of tubes and pipes to provide optimum surface for heat transfer while facilitating fluid transport. There is often the need to pack into small volumes as much tube or pipe surface area as possible. Two general configurations are possible namely random packing and ordered packing.

Since manufacturing processes, to be efficient, require order and repeatability, random stacking of pipes and tubes in convective heat transfer is a non-starter. We are left with ordered stacking of which two arrangements are most popular. These two, looking from the end of the tubes, are the square pitch and triangular pitch arrangements. These are illustrated in Figure 2.3.

If fluid flows across this stack of pipes or tubes, what dimension do we use to calculate the mean fluid velocity across the bank of tubes or pipes? What diameter do we use to estimate the surface area for convective heat transfer, especially, if we are to use correlations for estimating the film heat transfer coefficients similar to those for internal flow?

Equivalent Diameter

The answer is the use of an equivalent diameter. It is defined as the diameter of an equivalent circular pipe or tube which has the same wetted area and flow cross section as the bank of tubes or pipes. This makes sense because the wetted area represents the area across which convective heat is transferred between fluids separated by solid surfaces while the the flow cross section represents the area through which the bulk fluid, gaining or losing heat, flows.

For example, a circular pipe of diameter D would have a wetted area per unit length of πD. Such a pipe would, also, have a flow cross sectional area or volume, per unit length, of $\pi D^2/4$. By our definition above, the equivalent diameter of this pipe, D_E would then be

$$\frac{Cross\ SectionalFlow\ Area}{Wetted\ Perimeter} = \frac{4\pi D^2}{4}\cdot\frac{1}{\pi D} = \frac{D}{4} \qquad (2.50)$$

This is not quite true. If, however, we multiply equation (2.50) by 4, we get an equivalent diameter, D_e, that is closer to reality. That is

$$D_e = \frac{4 \, x \, Cross \, SectionalFlow \, Area}{Wetted \, Perimeter} \qquad (2.51)$$

We can, now, apply this to a rectangular pipe of cross sectional dimensions, H, for height and W for width. By equation (2.51), the equivalent diameter for this rectangular pipe is

$$D_e = \frac{4 \, x \, Cross \, SectionalFlow \, Area}{Wetted \, Perimeter} = \frac{4 \, x \, H \, x \, W}{2(H+W)} = \frac{2HW}{H+W} \qquad (2.52)$$

For a triangle for which the side of length a, is opposite the angle, a, side of length b is opposite the angle, β and side of length c, is opposite the angle, γ, such that

$$\frac{a}{\sin \alpha} = \frac{b}{\sin \beta} = \frac{c}{\sin \gamma} \qquad (2.53)$$

the equivalent diameter is

$$D_e = \frac{4 \, x \, Cross \, SectionalFlow \, Area}{Wetted \, Perimeter} = \frac{ab\sin \gamma}{2(a+b+c)} \qquad (2.54)$$

We can now apply this to the two popular arrangements of banks of tubes, namely the square pitch and the triangular pitch arrangements shown in Figure 2.3 for pipes each of outside diameter, D_0.

Fig. 2.3: Square and Triangular Pitch Arrangements

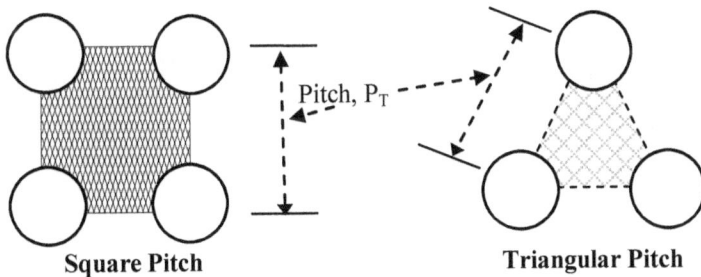

Square Pitch   **Triangular Pitch**

It can be seen that the equivalent diameters for the square and triangular pitch arrangements are, for the square pitch,

$$D_e = \frac{4 \, x \, Cross \, SectionalFlow \, Area}{Wetted \, Perimeter} = \frac{4x\left(P_T^2 - 4x\frac{1}{4}x\frac{\pi D_0^2}{4}\right)}{4x\frac{1}{4}x \pi D_0}$$

$$= \frac{4}{\pi D_0}\left(P_T^2 - \frac{\pi D_0^2}{4}\right) = \frac{0.273P_T^2}{D_0} - D_0 \qquad (2.55)$$

49

and for the triangular pitch, assuming an equilateral triangle,

$$D_e = \frac{4 \, x \, Cross \, SectionalFlow \, Area}{Wetted \, Perimeter} = \frac{4 \, x \left(\frac{P_T^2}{2} \sin 60 - 3 \, x \frac{1}{3} \, x \frac{\pi D_0^2}{4} \right)}{3 \, x \frac{1}{3} \, x \pi \, D_0}$$

$$= \frac{4}{\pi \, D_0} \left(0.433 P_T^2 - \frac{\pi D_0^2}{4} \right) = \frac{0.551 P_T^2}{D_0} - D_0 \qquad (2.56)$$

With the equivalent diameters determined, Colburn and Nusselt type correlations, used to determine the film heat transfer coefficient for flow across tube banks, are

i. Nusselt Type (*1000 < Re < 10,000*)

$$\frac{h_0 \, D_e}{k} = 0.36 \left(\frac{D_e \, G_s}{\mu} \right)^{0.55} \left(\frac{Cp \, \mu}{k} \right)^{\frac{1}{3}} \left(\frac{\mu_m}{\mu_s} \right)^{0.14} \qquad (2.57)$$

G_s is the mass flow outside the tubes per unit time per unit area.

ii. The Colburn Type

$$j_h = \frac{0.641}{Re^{0.54}} \qquad 10 < Re < 200 \qquad (2.58)$$

$$j_h = \frac{0.491}{Re^{0.49}} \qquad 200 < Re < 5,000 \qquad (2.59)$$

$$j_h = \frac{0.351}{Re^{0.45}} \qquad 5,000 < Re < 10^6 \qquad (2.60)$$

Bulk Flow Methods

In the design and analysis of heat exchangers, the desire is to maximize surface area while minimizing the pressure drop because high surface area means high rates of heat transfer while low pressure drop means low pumping costs. In what has come to be termed bulk flow methods, film heat transfer coefficients, in turbulent external flow, are estimated from empirical stream flow data obtained in heat exchanger-like configurations and compared to those obtained for cross flow over ideal tube banks.

Coulson et al (1983) report that, experimental work, for $1 < Re < 1000$ by Bergelin, Colburn and Hull and for $1000 < Re < 10^4$, by Bergelin, Brown and Doberstein, respectively, produced graphs from which the friction factor, f, can be determined. An example is the one shown in Figure 2.4 for square and triangular pitch arrangements. The friction factor, so determined, can be used with its corresponding Reynolds number, in the Colburn or Nusselt type correlation to estimate the film heat transfer coefficient.

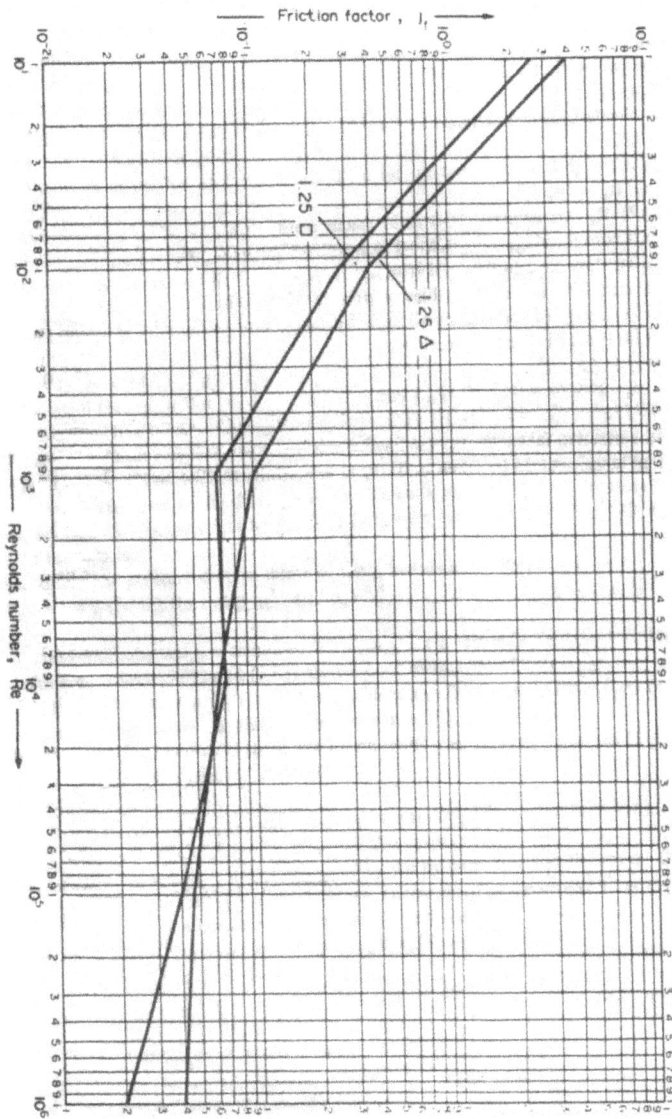

Fig. 2.4: Friction Factor vs Reynolds Number (Welty, 1978)

2.2.3: Film Heat Transfer Coefficient with Phase Change

The phase change we are considering here is that associated with boiling and condensation. Although freezing/thawing, sublimation and various other phase transformations, which result in different forms of the same substance such as ice, are also phase changes and are commercially important, our analysis here will be restricted to boiling and condensation.

Boiling Heat Transfer

Boiling and condensation involve very large heat fluxes on quite small temperature differences.The heat transfer coefficient, h, is, therefore expressed as a function of temperature difference. Recall that in heat transfer without phase change the heat transfer coefficient is expressed as a function of the fluid film thickness, δ, and its thermal conductivity, k, the fluid film thickness, δ, being a function of the Reynolds number in the system.

In boiling heat transfer the phase change sequence is:

$$SaturatedLiquid \rightarrow Vapour \rightarrow Gas \qquad (2.61)$$

When boiling occurs on a heated surface, submerged in a liquid at rest, it is known as pool boiling. When boiling occurs in a fluid moving past a hot surface, it is known as flow boiling. In each case, the temperature of the hot surface is a few degrees above the saturation temperature of the liquid. A section of a typical configuration of fluid boiling on a solid surface is shown in Figure 2.5

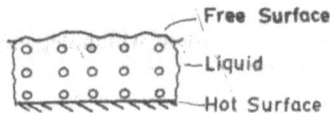

Fig. 2.5: Liquid Boiling on a Hot Surface

Experimental measurements, for pool boiling of water on the surface of a horizontal wire, have shown that, during boiling, the heat flux is, typically, related to temperature difference as shown in Figure 2.6. While the shape of the curve of Figure 2.6 is common to all liquids boiling on hot, solid, surfaces, the slopes, peaks and troughs of the curve may, however, vary for different liquid, surface and boiling situations.

At the time when the curve of Figure 2.6 was developed, computing and mathematical analysis had not advanced to the stage they are in today.

It is understandable; therefore, that this curve was divided into six sections in order to be able to isolate regimes in which workable mathematical relationships could be formulated between the heat flux and the temperature difference. It was beneficial, also, in directing attention to the regime of boiling most desirable for a given operation.

Fig. 2.6: Heat Flux versus Temperature Difference in Pool Boiling of Water on a Horizontal Wire (Welty, 1978)

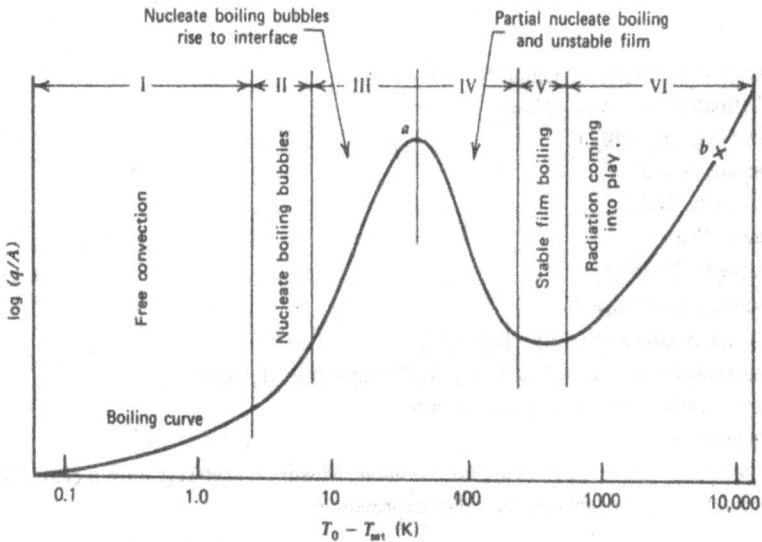

In regime I, for example, natural convection is the major energy exchange mechanism. In regime II, nucleate boiling and bubble formation predominate. Bubbles form on surface, reach a size sufficient for bouyancy forces to overcome surface tension forces, break off, rise through the cooler liquid and condense before reaching the free liquid surface. Bubbles form on favoured sites called active sites or nucleation sites.

In regime III, nucleate boiling proper is dominant. Bubbles are now able to reach the free liquid surface to expel vapour. Very high heat transfer rates, due mainly to the circulation of liquid past the hot surface not covered by vapour, are encountered. A stage is reached when enough of hot surface is covered with vapour that heat flux is decreased. Such a

53

point, of maximum heat flux, is called the BURNOUT point (for water at 1 atm, burnout occurs at $\Delta T \approx 55K$ and $q \approx 1.58 \text{ MW/m}^2$).

Regime IV is the transition boiling regime. More of the surface is covered by a film of vapour. Heat transfer is by conduction and convection through this film with decreasing amounts of surface being exposed directly to saturated liquid. Both nucleate and film boiling occur together.

Regime V is the stable film boiling regime and is characterised by the occurrence of a minimum in the q vs ΔT curve. In water, at atmospheric pressure, this occurs at $\Delta T \approx 280K$.

In regime VI, film boiling still persists but with so large a ΔT that radiation effects begin to dominate heat transfer. The heat flux begins to rise again with ΔT.

2.2.3.1: Correlations for Heat Transfer Coefficients in Phase Change

Because boiling heat transfer coefficients depend so much on the condition and location on the surface, it is difficult to predict them theoretically. However, several empirical correlations have been satisfactorily employed in a number of situations.

In stable nucleate pool boiling, Rohsenow (1952), in Welty (1978), found that

$$\frac{Cp_L \left(T_{surface} - T_{saturation}\right)}{h_{nb}} = C_{sf}\left(\frac{q}{\mu_L \, h_{nb}}\sqrt{\frac{\sigma}{g\left(\rho_L - \rho_v\right)}}\right)^{\frac{1}{3}}.\text{Pr}_L^{1.7} \quad (2.62)$$

where the subscripts, L and V refer to liquid and vapour and C_{sf} is an empirical constant which depends on the fluid and surface combination.

At atmospheric pressure, $T_{saturation}$ is the normal boiling point but, at other pressures, has to be the saturation temperature at that pressure. C_{sf} values given by Welty (1978) are shown below.

Fluid/Surface Combination	C_{sf}
water/copper	0.013
water/brass	0.006

When experimental values of C_{sf} are not available, Forster and Zubber

(1955) suggest the use of, (Coulson et al, 1983),

$$h_{nb} = 0.00122\left[\frac{k_L^{0.79}\,Cp_L^{0.45}\,\rho_L^{0.49}}{\sigma^{0.5}\,\mu^{0.29}\,\lambda^{0.24}\,\rho_v^{0.24}}\left(T_w - T_{satn}\right)^{0.24}\left(P_w - P_{satn}\right)^{0.75}\right] \quad (2.63)$$

where

h_{nb} = nucleate, pool boiling coefficient, $W/m^2.C$
k_L = liquid thermal conductivity, W/mC
Cp_L = liquid heat capacity, J/kg C
ρ_L = liquid density, kg/m^3
μ_L = liquid viscosity, Ns/m^2
λ = Latent heat, J/kg
ρ_v = vapour density, kg/m^3
T_w = wall, surface temperature, C
T_{satn} = saturation temperature of boiling liquid, C,
P_w = saturation pressure corresponding to the wall temperature T_w, N/m^2
P_{satn} = saturation pressure corresponding to T_{satn}, N/m^2
σ = surface tension, N/m.

When data on the physical properties of the fluid are not available, Mostinski (1963) gave an expression (Coulson et al, 1983) which predicts *h_{nb}* as reliably as the other expressions. This is

$$h_{nb} = 0.104P_c^{0.69}\,q^{0.7}\left[1.8\left(\frac{P}{P_c}\right)^{0.17} + 4\left(\frac{P}{P_c}\right)^{1.2} + 10\left(\frac{P}{P_c}\right)^{10}\right] \quad (2.64)$$

where P = operating pressure, bar
P_c = critical pressure of liquid, bar
q = h_{nb} (T_w - T_{satn}) = heat flux, W/m^2

The Maximum Heat Flux

To determine the maximum heat flux, as indicated in Figure 2.6, Rohsenow & Griffith (1955) give the expression, (Welty 1978),

$$\frac{q_{max}}{\rho_v h_{nb}} = 451g^{\frac{1}{4}}\left(\frac{\rho_L - \rho_v}{\rho_v}\right)^{0.6} \quad (2.65)$$

where *g* is the gravitational acceleration in G's and *q_{max}* has units of W/m^2. Zuber & Tribus propose in *fps* units, (Welty, 1978)

$$q_{max} = \left(\frac{\pi\lambda\rho_v}{24}\right)\left[g\sigma\left(\frac{\rho_L - \rho_v}{\rho_v^2}\right)\right]^{\frac{1}{4}}\left(\frac{\rho_L}{\rho_L + \rho_v}\right)^{\frac{1}{2}} \quad (2.66)$$

where

$$\lambda = \text{latent heat of vapour, Btu/lb}$$
$$\sigma = \text{surface tension, poundal/ft}$$
$$g = \text{gravitational acceleration, 32.174 ft/s}^2$$
$$\rho_L = \text{liquid density, lb/ft}^3$$
$$\rho_v = \text{density of vapour, lb/ft}^3$$

Mostinski (Coulson et al, 1983) also gives a reduced pressure equation, in S.I. units,

$$q_{max} = 3.67x10^4 \, P_c \left(\frac{P}{P_c}\right)^{0.35} \left(1 - \frac{P}{P_c}\right)^{0.9} \tag{2.67}$$

In stable film pool boiling, Bromley (1950) gives, for horizontal tubes, (Welty 1978),

$$h_{fb} = 3.52 \left[\frac{k_v \, \rho_v \, (\rho_L - \rho_v)g\,(\lambda + 0.4Cp_v \, \Delta T\,)}{D_o \, \mu_v (T_o - T_{satn})}\right]^{\frac{1}{4}} \tag{2.68}$$

where h_{fb} = heat transfer coefficient for film boiling
$\Delta T = T_o - T_{satn}$
D_o = outside diameter of tube
T_o = Temperature on the outside diameter of the tube.

Another expression, said to be in S.I. units, is

$$h_{fb} = 0.62 \left[\frac{k_v^3 \, \rho_v \, (\rho_L - \rho_v)g\,\lambda}{D_o \, \mu_v (T_w - T_{satn})}\right]^{\frac{1}{4}} \tag{2.69}$$

T_w is the mean tube wall temperature.

For a horizontal plane surface, Berenson (1969), as reported by Welty (1978), proposed replacing D_o, in either of equations (2.68) and (2.69), by

$$\left(\frac{\sigma}{g(\rho_L - \rho_v)}\right)^{\frac{1}{2}}$$ and evaluating k_v, ρ_v and μ_v at film temperature.

For vertical tubes, Hsu and Westwater (1958) give, (Welty 1978),

$$h_{fb} = \frac{0.0114Re^{0.6}}{\left(\dfrac{\mu_v^2}{g\rho_v(\rho_L - \rho_v)k_v^3}\right)^{\frac{1}{3}}} \tag{2.70}$$

where $\mathrm{Re} = \dfrac{4\,w}{\pi\,D_o\,\mu_v}$ and w is the vapour flowrate at the upper end of the tube. For boiling combined with convection

$$q_{total} = q_{convection} + q_{boiling} \qquad (2.71)$$

Condensation Heat Transfer

Condensation, on a surface, occurs when the surface is maintained at a temperature below the saturation temperature of an adjacent vapour.

<u>Filmwise condensation</u> occurs when liquid condensate wets the surface, spreads out and forms a film over the entire surface so that the vapour no longer makes contact, directly, with the surface. Heat transfer, in such a situation, is by conduction through the liquid film which constitutes, thereby, a resistance to heat transfer. The heat transfer rate (condensation rate) decreases, therefore, with increasing thickness of film.

<u>Dropwise condensation</u> occurs when a portion of the cold surface is not covered by condensate film and is in contact with the vapour and, therefore, not subject to the insulating effect of the liquid layer. Dropwise condensation is associated with higher heat transfer rates than filmwise condensation but is difficult to achieve and maintain for any extended length of time. Efforts have been made to use promoters on the surface to improve predictability of performance and/or efficiency of condensation. The most successful have been the use of silicone and Teflon® coated surfaces although these are expensive. Equipment design is, therefore, on the basis of film wise condensation and, hence, conservative.

Filmwise Condensation

For film condensation on a vertical wall, of length, L, inclined at an angle, φ, to the horizontal, Nusselt (Welty 1978) derived theoretical expressions for condensation on an inclined flat surface such that, in a performance situation

$$h_c = 0.943 \left(\frac{\lambda \rho^2 \, g \, k^3 \sin \varphi}{\mu L \left(T_i - T_w \right)} \right)^{\frac{1}{4}} \qquad (2.72)$$

and in a design situation

$$h_c = 0.925k \left(\frac{\rho^2 g \sin \varphi}{\mu F} \right)^{\frac{1}{3}}$$ (2.73)

where F is the mass flow rate of condensate. Because the Nusselt equations were found to give results that were 20% lower than observed values, Kirkbride and Badger (in McAdams, 1954) proposed the following corrected forms of the Nusselt's equations. In laminar flow

$$\frac{h}{k} \left(\frac{\mu^2}{\rho^2 g \sin \varphi} \right)^{\frac{1}{3}} = 1.11 \left(\frac{F_L}{\mu} \right)^{-\frac{1}{3}}$$ (2.74)

In turbulent flow

$$\frac{h}{k} \left(\frac{\mu^2}{\rho^2 g \sin \varphi} \right)^{\frac{1}{3}} = 0.0134 \left(\frac{F_L}{\mu} \right)^{0.4}$$ (2.75)

Equations (2.74) and (2.75), though derived for condensation on vertical planes, may be used for condensation outside vertical tubes if *sin φ = 1*.

For condensation on a bank of horizontal tubes at low vapour velocities

$$h = 0.725 \left(\frac{\lambda \rho^2 g k^3}{\mu N D_o (T_i - T_w)} \right)^{\frac{1}{4}} = 0.95 \left(\frac{\rho^2 g k^3 L}{\mu F_T} \right)^{\frac{1}{3}}$$ (2.76)

where

N	= number of tubes in a vertical row
D_o	= outside diameter of the tubes
L	= length of tube
F_T	= total condensation on the bottom tube (F = F_T/L)

When the effects of vapour velocity and density are taken into account, having been considered negligible in the theoretical derivations, the following version of the Nusselt equations, for film condensation on a vertical wall, emerges

$$h_L = 0.943 \left[\frac{\rho_L g k_L^3 (\rho_L - \rho_v) \{\lambda + \frac{3}{8} Cp_L (T_{satn} - T_{surface})\}}{L \mu (T_{satn} - T_{surface})} \right]^{\frac{1}{4}}$$ (2.77)

If Pr > 0.5 and $\dfrac{Cp_L (T_{satn} - T_{surface})}{\lambda} < 1.0$, 3/8 is replaced by 0.68.

Rohsenow obtained, for a plane wall at angle θ from the horizontal

$$h_L = 0.943 \left[\frac{\rho_L \, g \sin\theta \, k_L^3 \, (\rho_L - \rho_v) \left\{ \lambda + 0.68 Cp_L \left(T_{satn} - T_{surface} \right) \right\}}{L \mu \left(T_{satn} - T_{surface} \right)} \right]^{\frac{1}{4}} \qquad (2.78)$$

For film condensation on single horizontal cylinders Welty (1978) gives

$$h_{horiz} = 0.725 \left[\frac{\rho_L \, g \, k_L^3 \, (\rho_L - \rho_v) \left\{ \lambda + \frac{3}{8} Cp_L \left(T_{satn} - T_{surface} \right) \right\}}{D \mu \left(T_{satn} - T_{surface} \right)} \right]^{\frac{1}{4}} \qquad (2.79)$$

where D is the diameter of the cylinder. Note that for a cylinder, dividing equation (2.75) by (2.77) gives

$$\frac{h_{vertical}}{h_{horizontal}} = \frac{0.943}{0.725} \left(\frac{D}{L} \right)^{\frac{1}{4}} = 1.3 \left(\frac{D}{L} \right)^{\frac{1}{4}} \qquad (2.80)$$

It may be seen, from equation (2.80), that $h_{vertical} = h_{horizontal}$, when L/D is equal to 2.86. This means that a tube, with L/D = 2.86, will have the same heat transfer coefficient regardless of horizontal or vertical orientation. As L/D increases, however, condensation on a horizontal tube will be associated with greater heat transfer.

For a vertical bank of **n** horizontal tubes, according to Nusselt (1916)

$$h_{avg} = 0.725 \left[\frac{\rho_L \, g \, k_L^3 \, (\rho_L - \rho_v) \left\{ \lambda + \frac{3}{8} Cp_L \left(T_{satn} - T_{surface} \right) \right\}}{n D \mu \left(T_{satn} - T_{surface} \right)} \right]^{\frac{1}{4}} \qquad (2.81)$$

where **n** is the number of banks of horizontal tubes and D is the outside diameter of tube. Chen (1961) developed the expression, valid for
$$\frac{Cp_L \, (T_{satn} - T_{surface})(n-1)}{\lambda} > 2$$

$$h_{avg} = 0.725 \left[1 + \frac{0.02 Cp_L \, (T_{satn} - T_{surface})(n-1)}{\lambda} \right]$$

$$x \left[\frac{\rho_L \, g \, k_L^3 \, (\rho_L - \rho_v) \left\{ \lambda + \frac{3}{8} Cp_L \left(T_{satn} - T_{surface} \right) \right\}}{n D \mu \left(T_{satn} - T_{surface} \right)} \right]^{\frac{1}{4}} \qquad (2.82)$$

Chen's correlation is said to agree more closely with experimental results than Nusselt's equation.

WORKED EXAMPLES

Example 2.1

The heat transfer coefficient for natural convection near a heated vertical hot plate of length, L, is given by Eckert and Jackson (Welty, 1978) as

$$Nu_L = 0.555(Gr.Pr)^{\frac{1}{4}} \quad \text{for } Gr.Pr < 10^9$$

$$Nu_L = 0.0210(Gr.Pr)^{\frac{2}{5}} \quad \text{for } Gr.Pr > 10^9$$

Obtain the heat transfer coefficient for natural convection in air at 30 C near a vertical stainless steel plate, 60cm high, maintained at 60 C.

Answer

From the equations given

$$Nu_L = \frac{hL}{k} = a(Gr.Pr)^b$$

where a = 0.555, b = ¼ if Gr.Pr < 10^9 and a = 0.0210, b = 2/5 if Gr.Pr > 10^9.

Since $Pr = \frac{Cp\,\mu}{k}$ and $Gr = \frac{\rho^2 \beta g L^3 \Delta T}{\mu^2}$, we can get, from standard

tables of physical properties, that, for air

Cp	= 1.0063 kJ/kg.K	β	= 0.0577 m³/m³.K
μ	= 18.464 x 10^{-6} Ns/m²	ρ	= 1.1769 kg/m³
k	= 2.6240 x 10^{-2} W/m.K		

That is

$$Pr = \frac{Cp\,\mu}{k} = \frac{1.0063x18.464x10^{-6}x1000}{2.6240x10^{-2}}, \frac{kJ}{kg.K}\frac{N.s}{m^2}\frac{m.K}{W}\frac{J}{kJ} = 0.708$$

$$Gr = \frac{\rho^2 \beta g L^3 \Delta T}{\mu^2}$$

$$= \frac{(1.1769)^2 x0.0577x9.81x(0.6)^3 x30}{(18.464x10^{-6})^2}, \frac{kg^2}{m^6}\cdot\frac{m^3}{m^3.K}\cdot\frac{m}{s^2}\cdot\frac{m^3}{1}\cdot\frac{K}{1}\cdot\frac{m^4}{N.^2s^2}$$

$$= 1.4902x10^{10}$$

This gives us that Gr.Pr = 1.4902 x 10^{10} x 0.708 = 1.055 x 10^{10} > 10^9. That is, a = 0.0210 and b = 2/5 from which we get that

$$Nu_L = \frac{hL}{k} = a(Gr.\Pr)^b = 0.0210x(1.055x10^{10})^{\frac{2}{5}} = 21455$$

Thus

$$h = 21455\frac{k}{L} = 21455x\frac{0.02624}{0.6}\frac{W}{m.K}.\frac{1}{m} = 9.38,\frac{W}{m^2.K} \qquad Ans.$$

Example 2.2

The correlations for heat transfer in natural convection within rectangular enclosures of height, H, and width, L, for one vertical wall heated and one vertical wall cooled is given be Pagnani (Welty, 1978) as follows:

Conduction Regime: $Gr_L < 2$ x 10^3; $Nu_L = 1$
Laminar Flow Regime: 2 x $10^3 \le Gr_L \le 2$ x 10^5:

$$Nu_L = 0.18Gr^{\frac{1}{4}}\left(\frac{H}{L}\right)^{-\frac{1}{9}}$$

Turbulent Flow Regime: 2 x $10^5 \le Gr_L \le 2$ x 10^7;

$$Nu_L = 0.065Gr^{\frac{1}{3}}\left(\frac{H}{L}\right)^{-\frac{1}{9}}$$

Estimate the heat transfer coefficient in the middle of each regime for water enclosed between two such rectangular enclosures of height 0.6m and width 0.1m with one face at 100C and the other at 30 C,.

Answer

Water properties are evaluated at the arithmetic mean temperature of the walls, that is at mean temperature = (100 + 30)/2 + 273 = 338 K. At this temperature and from standard tables of properties:
    $\rho = 982$ kg/m^3 k = 0.663 W/m.K
    $\beta = 5.04$ x 10^{-4}, K^{-1} $\mu = 0.432$ x 10^{-3} Ns/m^2
In the conduction regime, the middle of the range is Gr = 2000/2 = 1000.
Since $Nu_L = 1$ anyway,

$$h_{conduction} = \frac{k}{L} = \frac{0.663}{0.6}\frac{W}{m.K}.\frac{1}{m} = 1.105,\frac{W}{m^2.K} \qquad Ans.$$

In the laminar flow regime, the middle of the range occurs at

$$Gr_{mean} = \frac{2x10^5 + 2x10^3}{2} = 1.01x10^5.$$

Thus

$$Nu_L = 0.18Gr^{\frac{1}{4}}\left(\frac{H}{L}\right)^{-\frac{1}{9}} = 0.18x\left(1.01x10^5\right)^{\frac{1}{4}} x\left(\frac{0.6}{0.1}\right)^{-\frac{1}{9}} = 2.6296$$

That is

$$h_{laminar} = 2.6296x\frac{k}{L} = 2.6296x\frac{0.663}{0.6}\frac{W}{m.K}.\frac{1}{m} = 2.906\frac{W}{m^2.K}\qquad Ans.$$

In the turbulent flow regime, the middle of the range is at

$$Gr_{mean} = \frac{2x10^5 + 2x10^7}{2} = 1.01x10^7.$$

This gives

$$Nu_L = 0.065Gr^{\frac{1}{3}}\left(\frac{H}{L}\right)^{-\frac{1}{9}} = 0.065x\left(1.01x10^7\right)^{\frac{1}{3}} x\left(\frac{0.6}{0.1}\right)^{-\frac{1}{9}} = 11.5140$$

That is

$$h_{turbulent} = 11.5140x\frac{k}{L} = 11.5140x\frac{0.663}{0.6}\frac{W}{m.K}.\frac{1}{m} = 12.723\frac{W}{m^2.K}\qquad Ans.$$

Thus, in this particular example, the heat transfer coefficient, in laminar, natural convection, is greater than that in the conduction regime i.e.

$$\frac{h_{laminar}}{h_{conduction}} = \frac{2.906}{1.105} = 2.63$$ while that in turbulent, natural convection is, also, greater than that in the conduction regime i.e.

$$\frac{h_{turbulent}}{h_{conduction}} = \frac{12.723}{1.105} = 11.51.$$ The comparable ratio for turbulent over laminar is $$\frac{h_{turbulent}}{h_{laminar}} = \frac{12.723}{2.906} = 4.38.$$ All these ratios illustrate the fairly large magnitude of the enhancement in heat transfer that occurs as we go from the conduction regime to the turbulent regime of natural convection.

Example 2.3

A cylindrical pipe, whose surface temperature is 422K, has an outside diameter of 0.10m. Air at 278 K blows across this pipe, normal to the pipe axis, at 7.6 m/s. Estimate the heat transfer rate per metre of pipe under these conditions.

Answer

Again, we estimate air properties at the arithmetic mean temperature of the bulk air and the surface temperature of the pipe, that is at mean temperature = (278 + 422)/2 = 350 K. At this temperature and from standard tables of properties, linear interpolation of air densities gives us

$$\frac{\rho - 1.0382}{0.9805 - 1.0382} = \frac{350 - 340}{360 - 340} = \frac{10}{20}$$

and

$$\rho = 1.0382 - 0.0577x\frac{10}{20} = 1.0094 \text{kg/m}^3.$$

Similarly for the air viscosity at this temperature.

$$\frac{\mu - 20.300}{21.175 - 20.300} = \frac{350 - 340}{360 - 340} = \frac{10}{20}$$

and

$$\mu = \left(20.300 + 0.875x\frac{10}{20}\right)x10^{-6} = 20.7375x10^{-6}, \text{Ns/m}^2$$

The thermal conductivity at the mean air temperature is, similarly,

$$\frac{k - 0.029282}{0.030779 - 0.029282} = \frac{350 - 340}{360 - 340} = \frac{10}{20}$$

and

$$k = 0.029282 + 0.001497x\frac{10}{20} = 0.0300, \text{W/m.K.}$$

We can assume, safely, that the equivalent diameter, for transverse air flow across the pipe diameter, is the pipe external diameter. Thus

$$Re_{air} = \frac{\rho D u}{\mu} = \frac{1.0094x0.10x7.6}{20.7375x10^{-6}}, \frac{kg}{m^3}.\frac{m}{1}.\frac{m}{s}.\frac{m^2}{N.s} = 36,993 \approx 3.7x10^4$$

This value of the Reynolds number indicates that flow is in forced convection and is in the range 4000 < Re_D < 40000 for which Nu_D = 0.174 $Re^{0.618}$. That is

$$Nu_D = 0.174Re^{0.618} = 0.174x(36993)^{0.618} = 115.78 \text{ and}$$

$$h = 115.78x\frac{k}{D} = 115.78x\frac{0.0300}{0.1}, \frac{W}{m.K}.\frac{1}{m} = 34.73, \frac{W}{m^2.K}$$

Since the external pipe area per metre is A = πD, the heat transfer rate per metre of pipe is

$$Q = hA\Delta T = 34.73x\pi x0.1x(422 - 278), \frac{W}{m^2.K}.\frac{m^2}{m}.\frac{K}{1} = 157178, \frac{W}{m}, \text{Ans.}$$

Example 2.4

It is desired to condense iso-amyl proprionate vapour in a shell and tube

heat exchanger, at its atmospheric boiling point of 160 C, using cooling water, entering at 21 C and leaving the exchanger at 48.9 C. The water flows inside the tubes. It is known that the heat transfer coefficient for condensation of iso-amyl proprionate vapour is $h_s = 782.4$ W/m².K.
A typical tube, in this exchanger, has an ID of 0.01580m and an OD of 0.01905 m, and is made of material whose thermal conductivity, k, is 377.4 W/m.K. Given that, for the water stream, Re = 21, 590 and that

$$j_h = \frac{0.027}{Re^{0.2}} = St.Pr^{\frac{2}{3}}.Vis^{0.14},$$ determine an overall heat transfer

coefficient for this typical pipe in the heat exchanger.

Answer

The overall heat transfer coefficient is given by

$$\frac{1}{U_i A_i} = \frac{1}{U_o A_o} = \frac{1}{h_i A_i} + \frac{x_w}{k_w A_w} + \frac{1}{h_o A_o} \tag{1}$$

where U is the overall heat transfer coefficient, h, the film heat transfer coefficient, A, the surface area, and x and k are the thickness and thermal conductivity of the tube material, respectively. The subscripts, i, w and o denote the inside, the wall and the shell-side of the tube, respectively. Since the heat transfer coefficient on the shell side is given, only that on the tube inner diameter side, h_i, needs to be determined. It is determined from the given correlation

$$j_h = \frac{0.027}{Re^{0.2}} = St.Pr^{\frac{2}{3}}.Vis^{0.14} \tag{2}$$

as follows. First of all, we determine the mean temperatures, at which fluid and material properties to be used to evaluate equation (2), would be evaluated. That is, for the water stream, the mean temperature is (21 + 48.9)/2 = 34.95 C and for the tube wall, the mean temperature is (34.95 + 160)/2 = 97.5 C.

The properties of water at the mean water temperature are, then, from standard tables,

Cp_i = 4.183 kJ/kg.K	k_i = 0.6259 x 10^{-3} kW/m.K
μ_i = 7.306 x 10^{-4} Ns/m²	ρ_i = 993.6 kg/m³

Water viscosity, evaluated at the mean tube wall temperature of 97.5 C, is $\mu_m = 3.06$ x 10^{-4} Ns/m². We can, now, calculate equation (2) after evaluating the Prandtl number and the viscosity correction factor as

$$Pr_i = \frac{Cp_i \, \mu_i}{k_i} = \frac{4183 x 0.0007306}{0.6259}, \frac{J}{kg.K} \cdot \frac{N.s}{m^2} \cdot \frac{m.K}{W} = 4.883 \qquad (3)$$

Viscosity correction factor $= \dfrac{\mu_i}{\mu_m} = \dfrac{7.306}{3.06} = 2.388 \qquad (4)$

Since $\quad Re_i = \dfrac{\rho_i \, D_i \, u_i}{\mu_i} = 21590$

$$u_i = \frac{21590 x \, \mu_i}{\rho_i \, D_i} = \frac{21590 x 0.0007306}{993.6 x 0.01580}, \frac{N.s}{m^2} \cdot \frac{m^3}{kg} \cdot \frac{1}{m} = 1.0048 \frac{m}{s} \qquad (5)$$

$$St = \frac{h_i}{\rho_i \, Cp_i \, u_i} = \frac{0.027}{Re^{0.2} \cdot Pr^{\frac{2}{3}} \cdot Vis^{0.14}} = . \frac{0.027}{(21590)^{0.2} \, (4.883)^{\frac{2}{3}} \, (2.388)^{0.14}}$$

$$= 0.001128$$

From equation (2)

$$h_i = 0.001128 \& \rho_i \, Cp_i \, u_i = 0.001128 \& 993.6 x 4183 x 1.0048 \frac{kg}{m^3} \cdot \frac{J}{kg.K} \cdot \frac{m}{s}$$

$$= 471073, \frac{W}{m^2 .K} \qquad (6)$$

We can, now, evaluate equation (1) for the overall heat transfer coefficient, based on the inside diameter of the tubes as

$$\frac{1}{U_i} = \frac{A_i}{U_o \, A_o} = \frac{1}{h_i} + \frac{x_w \, A_i}{k_w \, A_w} + \frac{A_i}{h_o \, A_o} \qquad (7)$$

and that based on the outside diameter of the tubes as

$$\frac{1}{U_o} = \frac{A_0}{U_i \, A_i} = \frac{A_o}{h_i \, A_i} + \frac{x_w \, A_o}{k_w \, A_w} + \frac{1}{h_o} \qquad (8)$$

Note that

$$\frac{A_i}{A_o} = \frac{\pi \, D_i}{\pi \, D_o} = \frac{D_i}{D_o} \qquad (9)$$

$$\frac{x_w \, A_i}{A_w} = \frac{(D_o - D_i) D_i}{(D_o + D_i)} .$$

$$\frac{x_w \, A_o}{A_w} = \frac{(D_o - D_i) D_o}{(D_o + D_i)} \qquad (9a)$$

Hence, from equations (7) and (9)

$$\frac{1}{U_i} = \frac{1}{471073} + \frac{(0.01905 - 0.01580) x 0.01580}{37.74 x \,(0.01905 + 0.01580)} + \frac{0.01580}{7824 x 0.01905}$$

$$= 0.0002122 \& 0.00003904 + 0.0010600 \overline{7} 0.00131139$$

That is $$U_i = 76255\frac{W}{m^2.K}, \; Ans \tag{10}$$

Similarly, from equations (8) and (9)

$$\frac{1}{U_o} = \frac{0.01905}{471073x0.01580} + \frac{(0.01905-0.01580)x0.01905}{37.74x(0.01905+0.01580)} + \frac{1}{7824}$$

$$= 0.00025595+ 0.00004707+ 0.00127812+ 0.00158114$$

That is

$$U_o = 63246\frac{W}{m^2.K}, \; Ans \tag{11}$$

Equations (10) and (11) show that the value, of the overall heat transfer coefficient, depends on the surface on which it is calculated.

Example 2.5

Ethylenc glycol, at 149 C, flows at 1.52 kg/s through a clean steel pipe having an outside diameter of 42.9 mm and a thickness of 0.003252 m. The atmospheric temperature is 27 C and the atmospheric film heat transfer coefficient may be taken as 11.36 W/m.K. The steel pipe is insulated with magnesia (85%) which is covered with 3.2 mm thickness of expanded cellular polystyrene which must be kept at 79 C. Determine the overall heat transfer coefficient, based on the inside diameter of the steel pipe, and the heat lost through the pipe. You are given that $j_H = St . Pr^{2/3} = 0.027 Re^{-0.2}$.

Data given are; for steel: $k_s = 45$ W/m.K, for magnesia, $k_m = 0.0366 + 8.72 x 10^5 T$, W/m.K, and for polystyrene: $k_p = - 0.0143 + 2.056 x 10^4 T$, W/m.K, where T is in degrees K. For ethylene glycol: $\mu = 9.5 x 10^{-4}$, Ns/m^2, $\rho = 1015$ kg/m^3, Cp = 3027 J/kg.K and $k_E = 0.1921$ W/m.K.

Answer

A cross section of the pipe may be represented as shown below.

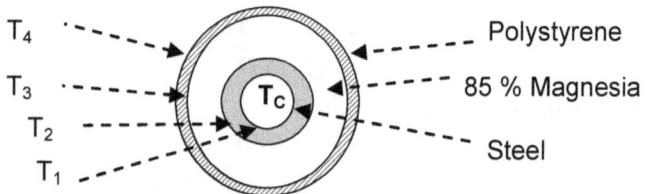

The first thing to do is to determine the diameters, D_1, D_2, D_3, and D_4, which correspond to the surface temperatures, T_1, T_2, T_3 and T_4. For the steel pipe, the outer diameter (OD) is D_2 and is 42.9 mm so that the inner diameter, (ID), D_1, is 42.9 - 2 x 3.252 = 36.396 mm. The other diameters are calculated similarly and all are summarised in the table below.

	Thickness, mm	ID, mm	OD, mm
Steel	3.252	$D_1 = 36.396$	$D_2 = 42.9$
85% Magnesia	unknown	$D_2 = 42.9$	$D_3 =$ unknown
Polystyrene	3.2	$D_3 =$ unknown	$D_4 = D_3 + 6.4$

The next thing is to determine the heat lost through each layer of material.

a. For flow inside the pipe

Heat is lost from the ethylene glycol inside the pipe. That is why, to minimise this heat loss, the pipe is insulated with magnesia and polystyrene. The heat transfer would, first of all, have to be through the thin boundary layer between the bulk of the ethylene glycol and the inner diameter of the steel pipe. This transfer would be associated with a heat transfer coefficient, h_i. Since we are given that $j_H = St .Pr^{2/3} = 0.027\ Re^{-0.2}$ we would need to obtain values for Re and Pr for flow inside the pipe in order to obtain the Stanton, and hence the heat transfer coefficient. Thus

$$mass\ flowrate,\ m = \rho u\ A = \frac{\rho u \pi D_1^2}{4} \qquad (1)$$

$$or \quad u = \frac{4m}{\rho \pi D_1^2} = \frac{4 x 1.52}{1015 x \pi x 0.0429^2}, \frac{kg}{s} \frac{m^3}{kg} \frac{1}{m^2} = 1.036 \frac{m}{s} \quad (2)$$

$$Re = \frac{\rho D_1 u}{\mu} = \frac{1015 x 0.0429 x 1.036}{9.5 x 10^{-4}}, \frac{kg}{m^3} \frac{m}{1} \frac{m}{s} \frac{m.s}{kg} = 4748533$$

$$Pr = \frac{Cp\ \mu}{k} = \frac{3027 x 9.5 x 10^{-4}}{0.1921}, \frac{J}{kg.K} \frac{kg}{m.s} \frac{s.m.K}{J} = 14.97 \quad (4)$$

Hence

$$St_i = \frac{h_i}{\rho_i\ Cp_i\ u_i} = \frac{0.027}{Re^{0.2}\ Pr^{\frac{2}{3}}} = \frac{0.027}{(4748533)^{0.2}\ x (14.97)^{\frac{2}{3}}} = 0.0005159$$

$or \quad h_i = 0.0005159 \ast \rho_i\, Cp_i\, u_i$ $\qquad$ (5)

$$= 0.0005159 \ast 1015 x 3027 x 1.036 \frac{kg}{m^3}.\frac{J}{kg.K}.\frac{m}{s} = 164215 \frac{W}{m^2.K}$$

We can, now, calculate the heat lost, per unit length of pipe, from the ethylene glycol inside the pipe, Q, as

$$Q = h_i\, A_i\, \Delta T_i = h_i\, \pi\, D_i\, (T_c - T_1) \quad or \quad T_c - T_1 = \frac{Q}{h_i\, \pi\, D_i} \quad (6)$$

For the heat lost, per unit length, through the walls of the steel pipe

$$Q = q.A = -k_s \frac{dT}{dr}.2\pi r \qquad (7)$$

A is the area normal to the direction of the heat flux, q, r is the radius of the pipe and Q is still the same. This equation can be rearranged and integrated so that

$$\int_{R_1}^{R_2} \frac{dr}{r} = -\frac{2\pi k_s}{Q} \int_{T_1}^{T_2} dT \quad from\,which \quad T_1 - T_2 = \frac{Q}{2\pi k_s} \ln\frac{R_2}{R_1} \quad (8)$$

For the heat lost, per unit length, through magnesia insulation

$$\int_{R_2}^{R_3} \frac{dr}{r} = -\frac{2\pi k_m}{Q} \int_{T_2}^{T_3} dT \quad from\,which \quad T_2 - T_3 = \frac{Q}{2\pi k_m} \ln\frac{R_3}{R_2} \quad (9)$$

Similarly, for the heat lost, per unit pipe length, through the polystyrene

Q is still the same and an equation similar to equation (8) applies. That is

$$\int_{R_3}^{R_4} \frac{dr}{r} = -\frac{2\pi k_p}{Q} \int_{T_3}^{T_4} dT \quad from\,which \quad T_3 - T_4 = \frac{Q}{2\pi k_p} \ln\frac{R_4}{R_3} \quad (10)$$

For the heat lost, per unit length, through the film of air

Since there is, also, a thin air film surrounding the polystyrene cover, the heat loss equation is similar to that for the ethylene glycol film. That is

$$Q = h_{air}\, A_{air}\, \Delta T_{air} = h_{air}\, \pi\, D_4\, (T_4 - T_{air}) \quad or \quad T_4 - T_{air} = \frac{Q}{h_{air}\, \pi\, D_4} \quad (11)$$

Adding equations (6) to (11), we get that

$$T_c - T_{air} = \frac{Q}{h_i\, \pi\, D_i} + \frac{Q}{2\pi k_s} \ln\frac{R_2}{R_1} + \frac{Q}{2\pi k_m} \ln\frac{R_3}{R_2}$$

$$+\frac{Q}{2\pi k_p}\ln\frac{R_4}{R_3}+\frac{Q}{h_{air}\,\pi\,D_4} \tag{12}$$

If U_i is the overall heat transfer coefficient based on the inside surface of the pipe, we know that

$$\frac{T_c-T_{air}}{Q}=\frac{1}{U_i\,A_i}=\frac{1}{h_i\,\pi\,D_i}+\frac{1}{2\pi k_s}\ln\frac{D_2}{D_1}+\frac{1}{2\pi k_m}\ln\frac{D_3}{D_2}$$

$$+\frac{1}{2\pi k_p}\ln\frac{D_4}{D_3}+\frac{1}{h_{air}\,\pi\,D_4} \tag{12a}$$

Now, $T_c=149+273=422$ K, $T_4=79+273=352$ K; $T_{air}=27+273=300$ K. At the mean temperature of $(422+352)/2=387$ K, the thermal conductivities of magnesia and polystyrene are, respectively,

$$k_m=0.0366+8.72x10^{-5}\,x\,387=0.0704\,\frac{W}{m.K}$$

$$k_p=-0.0143+2.056x10^{-4}\,x\,387=0.0653\,\frac{W}{m.K} \tag{13}$$

The only unknowns in equation (12) are Q and D_3. We need two equations, therefore, to solve for them. The best equations to use are equations (11) and (12) because they contain the greatest number of known variables in the problem. Thus, from equation (11) and the data given

$$Q=h_{air}\,\pi\,D_4\left(T_4-T_{air}\right)=11.36x\,\pi\,x\left(D_3+0.0032\right)\left(352-300\right),\frac{W}{m^2K}.\frac{m^2}{m}.\frac{K}{1}$$

$$=186680D_3+5.93856516\frac{W}{m} \tag{14}$$

From equation (12a) and the data given

$$\frac{422-300}{Q}=\frac{1}{164215\pi\,x0.036396}+\frac{1}{2\pi\,x45}\ln\frac{0.0429}{0.036396}+\frac{1}{2\pi\,x0.0704}\ln\frac{D_3}{0.0429}$$

$$+\frac{1}{2\pi\,x0.0653}\ln\frac{(D_3+0.0032)}{D_3}+\frac{1}{11.36\pi\,(D_3+0.0032)} \tag{15}$$

That is

$$\frac{122}{Q}=5.32364244x10^{-3}+5.81257895x10^{-4}+2.25981405\ln\frac{D_3}{0.0429}$$

$$+2.43630795\ln\frac{(D_3+0.0032)}{D_3}+\frac{0.02800896}{(D_3+0.0032)} \tag{16}$$

By labelling Q in equation (14) as Q_1 and that in equation (16) as Q_2, we can compute each for various values of D_3 as shown shown in the Table below.

Table Showing Computed Values of Q_1 and Q_2

D_3	Q_1	Q_2
0.048	95.54497	126.5516
0.050	99.27857	118.4904
0.051	101.1454	114.9013
0.052	103.0122	111.565
0.053	104.879	108.4557
0.054	106.7458	105.5509
0.055	108.6126	102.8309
0.056	110.4794	100.2787

When these values are plotted, as in the Figure shown, we can see that the D_3 at which $Q_1 = Q_2 = Q$ is 0.053746m.

Graphical Plot of the Computed Values of Q_1 and Q_2

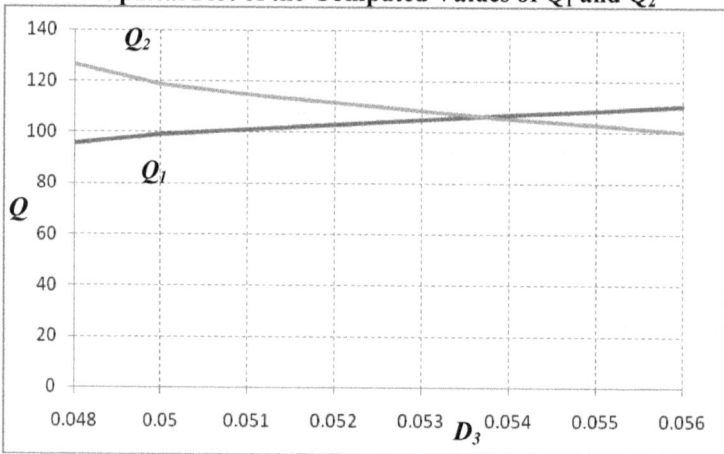

We can, now, calculate Q as 106.27 W. From equation (12a) we can see that

$$\frac{T_c - T_{air}}{Q} = \frac{1}{U_i A_i} \quad or \quad U_i = \frac{Q}{\pi D_i (T_c - T_{air})}, \frac{J}{s}.\frac{1}{m^2}.\frac{1}{K}$$

$$= \frac{106.27}{\pi \times 0.036396(422 - 300)} = 167.53, \frac{W}{m^2.K} \quad Ans$$

Example 2.6

It is desired to evaporate 30 kg/h of water, at 1 atm, by passing stream through a vertical tube 19 mm diameter in a pool of water. Estimate the

70

boiling heat transfer coefficient on the outside of the tube given that this heat transfer coefficient, h, is given by Tsu and Westwater (1958) as

$$h = \frac{0.0114Re^{0.6}}{\left(\dfrac{\mu_v^2}{g\rho_v(\rho_L - \rho_v)k_v^3}\right)^{\frac{1}{3}}}$$

where

$$Re = \frac{4w}{\pi D_o \mu_v}$$

w = Vapour rate at the top of the tube, kg/s
D_o = outside diameter of the tube, m
μ_v = viscosity of vapour, Ns/m^2
ρ_L, = density of liquid, kg/m^3
ρ_v = density of vapour, kg/m^3
k_v = thermal conductivity of vapour, W/m.K
g = acceleration due to gravity, m/s^2

You are provided with steam tables.

Answer

Given $w = 30\dfrac{kg}{h}.\dfrac{h}{3600s} = 8.33x10^{-3},\dfrac{kg}{s}$, D_o = 0.019 m

From steam tables: at 100 C, 1 atm
μ_v = 12 x 10^{-6}, Ns/m^2
ρ_v = 1/1.673 = 0.5977, kg/m^3
$k_v = 24.8x10^{-6}\dfrac{kW}{m.K}.\dfrac{1000W}{kW} = 24.8x10^{-3},\dfrac{W}{m.K}$

$\rho_L = \dfrac{1}{0.001044} = 957.85,\dfrac{kg}{m^3}$

$w = 30\dfrac{kg}{h}.\dfrac{h}{3600s} = 8.33x10^{-3},\dfrac{kg}{s}$

Then

$$Re = \frac{4w}{\pi D_o \mu_v} = \frac{4x8.33x10^{-3}}{\pi x0.019x12x10^{-6}},\frac{kg}{s}.\frac{1}{m}.\frac{m.s}{kg} = 46499$$

These enable us to evaluate h as

$$h = \frac{0.0114 Re^{0.6}}{\left(\dfrac{\mu_v^2}{g\,\rho_v\,(\rho_L - \rho_v)k_v^3}\right)^{\frac{1}{3}}}, \quad \frac{1}{\left(\dfrac{kg^2}{m^2.s^2} \cdot \dfrac{s^2}{m} \cdot \dfrac{m^3}{kg} \cdot \dfrac{m^3}{kg} \cdot \dfrac{m^3.K^3}{W^3}\right)^{\frac{1}{3}}}$$

$$= \frac{0.0114 x (46499)^{0.6}}{\left(\dfrac{\left(12 x 10^{-6}\right)^2}{9.81 x 0.5977 x (957.85 - 0.5977) x \left(24.8 x 10^{-3}\right)^3}\right)^{\frac{1}{3}}}$$

$$= \frac{7.2006}{\left(\dfrac{1.44 x 10^{-10}}{0.08561181}\right)^{\frac{1}{3}}} = \frac{7.2006}{1.18925852} = 60547, \frac{W}{m^2.K} \quad Ans.$$

Example 2.7

Iso - amyl propionate vapour is to be condensed at its atmospheric boiling point of 160 C in a horizontal shell and tube heat exchanger. The coolant is water which enters at 21 C and leaves at 48.9 C. The mass flowrate of water through the inside of the tube is 987.6 kg/s.m². Calculate the inside film heat transfer coefficient given that

$$j_h = \frac{0.027}{Re^{0.2}} = St.Pr^{\frac{2}{3}} . \left(\frac{\mu}{\mu_m}\right)^{0.14}$$

and that, for the tubes, D_o = 19 mm and D_i = 15.875 mm.

Answer

Mean water temperature = 273 + (21 + 48.9)/2 = 307.95 K
From standard tables, at 307.95 K, and for water,
μ = 7.19 x 10^{-4}, N.s/m², ρ = 994.1 kg/m³, Cp = 4.178 kJ/kg.K

$$\frac{k - 0.61}{0.62 - 0.61} = \frac{307.95 - 300}{311 - 300}$$

That is : $k = 0.61 + 0.01 x 0.7227 = 0.617, \dfrac{W}{m.K}$

Mean tube wall temperature = 273 + {160 + (21 + 48.9)/2}/2 = 370.48 K
At this temperature, and from standard tables,

$$\frac{\mu_m - 0.000297}{0.000281 - 0.000297} = \frac{370.48 - 368}{373 - 368}$$

or $\mu_m = 0.000297 - 0.000016 \times 0.496 = 2.89064 \times 10^{-4}, \dfrac{N.s}{m^2}$

$Vis \, cosity \, ratio = \dfrac{\mu}{\mu_m} = \dfrac{7.19 \times 10^{-4}}{2.89 \times 10^{-4}} = 2.49$

$Pr = \dfrac{Cp \, \mu}{k} = \dfrac{4178 \times 0.000719}{0.617} \dfrac{J}{kg.K} \cdot \dfrac{kg}{m.s} \cdot \dfrac{s.m.K}{J} = 4.87$

To estimate the Reynolds number, recall that

$Re_i = \dfrac{\rho_i \, u_i \, D_i}{\mu_i} = \dfrac{987.6 \times 0.015875}{0.000719} \cdot \dfrac{kg}{s.m^2} \cdot \dfrac{m}{1} \cdot \dfrac{m.s}{kg} = 218055$

We can, now, evaluate the Stanton number from

$$j_h = \frac{0.027}{Re^{0.2}} = St.Pr^{\frac{2}{3}} \cdot \left(\frac{\mu}{\mu_m}\right)^{0.14}$$

as

$$St_i = \frac{h_i}{\rho_i \, Cp_i \, u_i} = \frac{0.027}{Re_i^{0.2} \cdot Pr_i^{\frac{2}{3}} \cdot \left(\dfrac{\mu_i}{\mu_m}\right)^{0.14}} = \frac{0.027}{(218055)^{0.2} \times (4.87)^{\frac{2}{3}} \times (2.49)^{0.14}}$$

From which we get that

$h_i = \dfrac{987.6 \times 4178 \times 0.027}{7.37417 \times 2.8731 \, 1 \times 1.13623} \cdot \dfrac{kg}{s.m^2} \cdot \dfrac{J}{kg.K} = 46279, \dfrac{W}{m^2.K}.$ $Ans.$

Example 2.8

The heat heat transfer coefficient, for film condensation on single vertical cylinders, when Pr > 0.5 and $\dfrac{Cp_L (T_{satn} - T_{surface})}{\lambda} < 1.0$, is given by Nusselt as

$$h = 0.943 \left[\frac{\rho_L \, g \, k_L^3 \, (\rho_L - \rho_v) \cdot \left\{\lambda + \dfrac{3}{8} Cp_L (T_{satn} - T_{surface})\right\}}{L.\mu (T_{satn} - T_{surface})}\right]^{\frac{1}{4}}$$

where L is the length of the cylinder. If saturated steam at 1 atm. and 373 K is condensing on an 0.6 m x 1.0 m vertical, rectangular surface at 297

K, determine the appropriate film heat transfer coefficient and the rate of condensation given that for

steam	T_{satn} = 373 K	water	ρ_L = 982 kg/m³
	ρ_v = 0.598 kg/m³		Cp_L = 4.187 kJ/kg.K
	λ = 2257 kJ/kg		μ_L = 5.0 x 10⁻⁴ N.s/m²
			k_L = 0.614 W/m.K

Answer

Since $Pr = \dfrac{Cp\,\mu}{k} = \dfrac{4187x0.0005}{0.614}, \dfrac{J}{kg.K}.\dfrac{kg}{m.s}.\dfrac{s.m.K}{J} = 3.41 > 0.5$

and

$\dfrac{Cp_L(T_{satn} - T_{surface})}{\lambda} = \dfrac{4187x(373-297)}{2257000}, \dfrac{J}{kg.K}.\dfrac{K}{1}.\dfrac{kg}{J} = 0.141 < 1.0$

we can use the given equation on the condition that an equivalent circular diameter for the given rectangular cross section is used. Thus

$$h = 0.943\left[\dfrac{982x9.81x(0.614)^3 \; x \; (982-0.598)\left\{2257000+\dfrac{3}{8}x4187(373-297)\right\}}{1x(0.0005)x(373-297)}\right]^{\frac{1}{4}}$$

$$= 0.943x\left(\dfrac{5.20042942\text{х}10^{12}}{0.038}\right)^{\frac{1}{4}}\left(\dfrac{kg}{m^3}.\dfrac{m}{s^2}.\dfrac{J^3}{s^3m^3K^3}.\dfrac{kg}{m^3}.\dfrac{J}{kg}.\dfrac{1}{m}.\dfrac{m.s}{kg}.\dfrac{1}{K}\right)^{\frac{1}{4}}$$

$= 34203, \dfrac{W}{m^2.K}.$ Ans.

$Equivalent diameter\; D_e = \dfrac{4x0.6x1.0}{2x(0.6+1.0)} = 0.75m$

$Rate\,of\,condensation = h\,x\,\pi\,x\,D_e\;x\,L = 34203\,x\,\pi\,x\,0.75x1.0, \dfrac{W}{m^2.K}.\dfrac{m}{1}.\dfrac{m}{1}$

$= 806214W.$ Ans

Example 2.9

Saturated steam at 370 K is condensed on a 19mm tube with a surface temperature of 345 K. Calculate *h* for such a tube 1.5 m long and oriented horizontally. You are given that

$$h = 0.725 \left[\frac{\rho_L \, g \, k_L^3 \, (\rho_L - \rho_v) . \left\{ \lambda + \frac{3}{8} Cp_L \left(T_{satn} - T_{surface} \right) \right\}}{D . \mu \left(T_{satn} - T_{surface} \right)} \right]^{\frac{1}{4}}$$

where D = diameter of cylinder. Subscript L refers to liquid, V to vapour. λ is the latent heat, ρ the density, Cp the heat capacity and μ the viscosity. You are also provided with a steam table.

Answer

Condensed liquid properties will be evaluated at the mean temperature of condensing steam and tube surface. Saturated steam properties will be evaluated at the saturated steam temperature. Thus, at 370 K, $\rho_v = $ 0.537 kg/m^3 and λ_{370} = 2265.2 kJ/kg.
The mean film temperature = (370 + 345)/2 = 357.5 K. At this temperature, the properties of condensate are, from standard tables:

ρ_L = 969.5 kg/m^3 μ_L = 0.349 x 10^{-3} N.s/m
k_L = 0.674 W/mK Cp_L = 4.215 kJ/kgK

Since $T_{satn} - T_{surface}$ = 25 K, g = 9.81 m/s^2 and D = 0.019m

$$h = 0.725 \left[\frac{969.5 \times 9.81 \times (0.674)^3 \times (969.5 - 0.537) \times \left\{ 2265200 + \frac{3}{8} \times 4215 \times 25 \right\}}{0.019 \times 0.00034 \times 25} \right]^{\frac{1}{4}}$$

$$= 0.725 x \left[\frac{6.503109254 \, 10^{12}}{0.000165775} \right]^{\frac{1}{4}} \left[\frac{kg}{m^3} . \frac{m}{s^2} . \frac{J^3}{s^3 . m^3 . K^3} . \frac{kg}{m^3} . \frac{J}{kg} . \frac{1}{m} . \frac{m.s}{kg} . \frac{1}{K} \right]^{\frac{1}{4}}$$

$$= 1020325, \frac{W}{m^2 . K}. \quad Ans.$$

Example 2.10

n - hexane is to be condensed on a single horizontal tube by pumping water, at 12 m/s, through a type 304 stainless steel tube, 13mm ID and 1.5mm wall thickness. The water enters at 30 C and leaves at 50 C. Determine the overall heat transfer coefficient based on the outside surface of the tube. You are given that for flow inside tubes.

$$St_i = \frac{h_i}{\rho_i \, Cp_i \, u_i} = \frac{0.027}{Re_i^{0.2} \cdot Pr_i^{\frac{2}{3}}}$$

and for condensation outside the tubes

$$h = 0.725 \left[\frac{\rho_L \, g \, k_L^3 \, (\rho_L - \rho_v) \cdot \left\{ \lambda + \frac{3}{8} Cp_L \left(T_{satn} - T_{surface} \right) \right\}}{D \cdot \mu \left(T_{satn} - T_{surface} \right)} \right]^{\frac{1}{4}}$$

where the symbols have their usual meaning. The boiling point of n-hexane at 1 atm is 69C. You are provided with steam tables and the relevant properties of n-hexane as follows

ρ_L = 654.8 kg/m³ ; λ = 365 kJ/kg
ρ_v = 3.38 kg/m³ ; k_L = 0.124 W/mK
Cp_L = 2,260 J/kgK ; μ_L = 2.97 x 10⁻⁴ Ns/m²

For stainless steel, k_s = 16.55 W/m.K. Note that g = 9.81 m/s².

Answer

For flow of water inside the tube

Mean water temperature is *(50 + 30)/2 = 40 C.*
Since the inside diameter, ID, of the tube is 0.013m, the outside diameter, OD, is then *0.013 + 2 x 0.0015 = 0.016 m.*

At 40 C, the relevant properties of water are: ρ = 995 kg/m³, Cp = 4183 J/kg K, μ = 0.682 x 10⁻³ Ns/m², k = 0.630 W/mK. Thus

$$Re_i = \frac{\rho_i \, u_i \, D_i}{\mu_i} = \frac{995 x 12 x 0.013}{0.000682}, \frac{kg}{m^3} \cdot \frac{m}{s} \cdot \frac{m}{1} \cdot \frac{m.s}{kg} = 2275953$$

$$Pr = \frac{Cp\,\mu}{k} = \frac{4183 x 0.000682}{0.630}, \frac{J}{kg.K} \cdot \frac{kg}{m.s} \cdot \frac{s.m.K}{J} = 4.53$$

$$St_i = \frac{h_i}{\rho_i \, Cp_i \, u_i} = \frac{0.027}{Re_i^{0.2} \cdot Pr_i^{\frac{2}{3}}} = \frac{0.027}{(2275953)^{0.2} x (4.53)^{\frac{2}{3}}} = 0.000836627859$$

$$h_i = 0.000836627859 x 995 x 4183 x 12, \frac{kg}{m^3} \cdot \frac{J}{kg.K} \cdot \frac{m}{s} = 417854, \frac{W}{m^2.K}$$

For flow outside the tube, n- hexane is the fluid involved

$$h_o = 0.725 \left[\frac{\rho_L \, g \, k_L^3 \, (\rho_L - \rho_v). \left\{ \lambda + \frac{3}{8} Cp_L \left(T_{satn} - T_{surface} \right) \right\}}{D.\mu \left(T_{satn} - T_{surface} \right)} \right]^{\frac{1}{4}}$$

That is, with $(\rho_L - \rho_v) = 654.8 - 3.38 = 651.42$ and
$$(T_{satn} - T_{surface}) = (69 + 273) - (40 + 273) = 29 K$$

$$h_o = 0.725 \left[\frac{\rho_L \, g \, k_L^3 \, (\rho_L - \rho_v). \left\{ \lambda + \frac{3}{8} Cp_L \left(T_{satn} - T_{surface} \right) \right\}}{D.\mu \left(T_{satn} - T_{surface} \right)} \right]^{\frac{1}{4}}$$

$$= 0.725 \left[\frac{654.8 \times 9.81 \times (0.124)^3 \times 651.42 \times \left\{ 365000 + \frac{3}{8} \times 2260 \times 29 \right\}}{0.016 \times 0.000297 \times 29} \right]^{\frac{1}{4}}$$

$$= 0.725 [217924] \left[\frac{kg}{m^3} \cdot \frac{m}{s^2} \cdot \frac{J^3}{s^3.m^3.K^3} \cdot \frac{kg}{m^3} \cdot \frac{J}{kg} \cdot \frac{1}{m} \cdot \frac{m.s}{kg} \cdot \frac{1}{K} \right]^{\frac{1}{4}}$$

$$= 157995, \frac{W}{m^2.K}$$

Since

$$\frac{1}{U_o} = \frac{A_0}{U_i A_i} = \frac{A_o}{h_i A_i} + \frac{x_w A_o}{k_w A_w} + \frac{1}{h_o} \tag{8}$$

and

$$\frac{A_i}{A_o} = \frac{\pi D_i}{\pi D_o} = \frac{D_i}{D_o} \tag{9}$$

$$\frac{x_w A_o}{A_w} = \frac{(D_o - D_i) D_o}{(D_o + D_i)} \qquad from\,(9a)$$

$$\frac{1}{U_o} = \frac{D_o}{h_i D_i} + \frac{(D_o - D_i) D_o}{k_w (D_o + D_i)} + \frac{1}{h_o}$$

$$= \frac{0.016}{417854 \, x \, 0.013} + \frac{(0.016 - 0.013) \, x \, 0.016}{16.55 \, x \, (0.016 + 0.013)} + \frac{1}{157995}, \frac{m^2 . K}{W}$$

$$= 0.29455 x 10^{-4} + 1.00010 x 10^{-4} + 6.3293 x 10^{-4}$$

$$= 7.62396 x 10^{-4}, \frac{m^2 . K}{W}$$

That is, $U_o = 1311.65$ W/m^2.K. Ans.

References for Chapter Two

1   Churchill S. W., (1977), *Friction Factor Equation Spans all Fluid Flow Regimes*, Chemical Engineering, Vol. 84, No. 24, pp 91 -92, McGraw - Hill Book Company, New York, U.S.A

2   Churchill S. W., (1977), Ind. Eng. Chem. Fundam., Vol. 16, No. 1, pp 109 - 116, American Chemical Society, Washington D.C., U.S.A.

3   Coulson J. M, Richardson J. F., and Sinnott R. K., Chemical Engineering Vol. 6, Design, Pergamon Press, Oxford, UK., 1983.

4   Perry R., Green D., Chemical Engineers' Handbook, 6th. Edition, McGraw - Hill Book Company, New York, U.S.A., 1984.

5   Welty J. R., Engineering Heat Transfer, SI Version, John Wiley & Sons, New York, U.S.A., 1978

6   Welty J. R., Wicks C. E., and Wilson R. E., Fundamentals of Momentum, Heat and Mass Transfer, 2nd. Edition. John Wiley & Sons, New York, U.S.A., 1976

7   Kays, William; Crawford, Michael; Weigand, Bernhard (2004). Convective Heat and Mass Transfer, 4E. McGraw-Hill Professional, New York, U.S.A

8   W. McCabe J. Smith (1956). Unit Operations of Chemical Engineering. McGraw-Hill Book Company, New York, U.S.A

9   Bennett (1962). Momentum, Heat and Mass Transfer. McGraw-Hill Book Company, New York, U.S.A

10   http://en.wikipedia.org/wiki/Natural convection

CHAPTER THREE
THE THEORY AND DESIGN OF HEAT EXCHANGERS

3.1: Heat Exchangers

A heat exchanger is a device for transfering heat from one fluid to another usually without mixing of the fluids although in some cases, the fluids may be in direct contact. Heat exchangers are widely used in space heating, refrigeration, air conditioning, power plants including automotive, marine and aircraft engines, chemical plants, petrochemical plants, petroleum refineries, and natural gas processing.

It is not difficult to imagine that, in these applications, there would be many possible configurations of heat exchanger surface and fluid flows depending on cost and suitability for the purpose for which the heat exchanger is desired. Fortunately, however, most designs and configurations of heat exchangers can be put in one or more of four classes of heat exchangers.

1. Direct Transfer Heat Exchangers

In this type, two or more fluids, exchanging heat energy directly, are separated by a heat transfer surface. Direct transfer heat exchangers should not be confused with direct contact heat exchangers in which the fluids exchanging heat are not separated by a solid surface. The Leibig condenser, commonly used in many Chemistry laboratories, is a simple but good example of a direct transfer heat exchanger.

Examples of direct contact heat exchangers are to be found in the transfer of heat between a gas and a liquid in the form of drops, films or sprays. Such types of heat exchangers are used predominantly in air conditioning, humidification, water cooling and condensing plants.

Typically, the hardware and fluid flow arrangements, for a direct transfer heat exchanger, are as shown, schematically, below.

Fluid 2 In

Fluid 1 In →

Fluid 1 Out

Fluid 2 Out

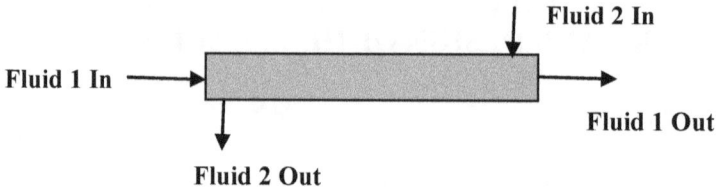

This schematic suggests that the fluid flow configuration can be any one of the following:

i   Co - current   - both fluids flow parallel along the same direction
ii   Counter current   - both fluids flow parallel but in opposite directions
iii Cross-flow   - both fluids flow at some angle (usually close to 90°) in either co-current or counter-current flow
iv Combined flow - both fluids flow in cross-flow and either co-current or counter-current manner.

You can imagine that there could be as many designs of direct transfer heat exchangers as there are heat exchange problems. Fortunately, a few standard designs have emerged from commercial practice and from their usefulness in various applications. These are the shell and tube heat exchanger (the work horse of the chemical and petroleum industries), the coiled pipe and plate heat exchangers (most commonly used in the food, pharmaceutical and domestic heating industries), heat pipes, double pipe heat exchangers, waste heat recuperators (used in miniaturized appliances, waste heat recovery and energy optimization). Even then, images of a few direct transfer heat exchangers, as advertised on their manufacturers' or suppliers' web sites, and shown below, illustrate the variety and complexity of designs available.

Figure 3.1: Shell and Tube Heat Exchanger (Diversified Heat Transfer Ltd)

Figure 3.2: Platular® Range **Plate Heat Exchanger (UK Exchangers)**

Figure 3.3: Coils of a Spiral Heat Exchanger (Polimer-Krakow)

Figure 3.4: Shell OEM Coiled Copper Heat Exchanger

Figure 3.5a: Econotherm Waste Heat Recuperators

Figure 3.5b: Econotherm Waste Heat Recuperators

2. Liquid Coupled Indirect Transfer Heat Exchangers

In this case, two direct transfer heat exchanger units are made to work together to effect the desired heat transfer by means of a heat transfer medium. This medium may be liquid or gas which collects heat from one unit and transfers it to the other. Typical examples of such heat exchangers are:

i   the automobile engine block and air cooled radiator in which the heat transfer medium is water
ii   the evaporator and condenser in a refrigerator or an air conditioner in which the heat transfer medium is a refrigerant.
iii the heat pipe, with hot and cold ends, in which the heat transfer medium is a fluid in natural convection

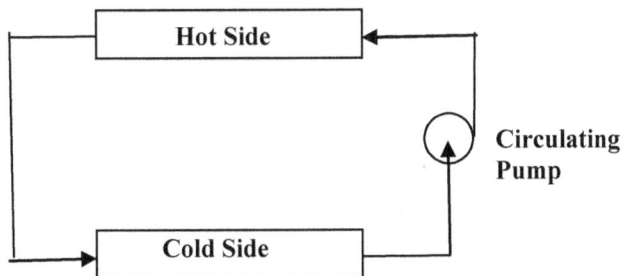

Figure 3.6: Liquid Coupled Indirect Transfer Heat Exchanger Schematic

Hot Reservoir

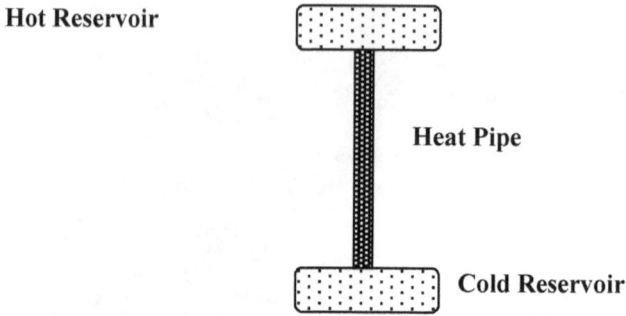

Heat Pipe

Cold Reservoir

Figure 3.7: Heat Pipe Heat Exchanger Schematic

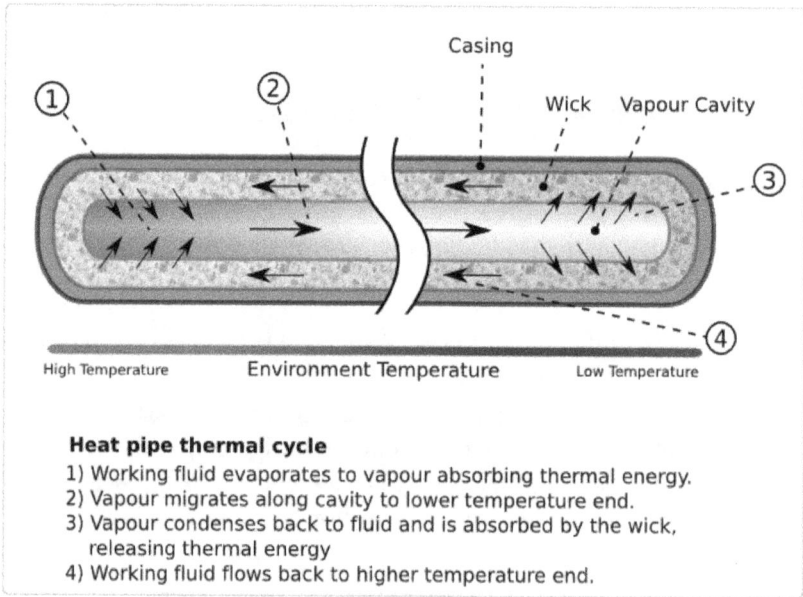

Heat pipe thermal cycle
1) Working fluid evaporates to vapour absorbing thermal energy.
2) Vapour migrates along cavity to lower temperature end.
3) Vapour condenses back to fluid and is absorbed by the wick, releasing thermal energy
4) Working fluid flows back to higher temperature end.

Figure 3.8: How the Heat Pipe Heat Exchanger Works (Wikipedia, 2010)

3. Periodic Flow Heat Exchangers

This time, a solid object, capable of conducting heat energy, rather than a fluid medium, serves as the medium of heat transfer between a hot and cold fluid. In gas turbine systems, for example, a metal matrix

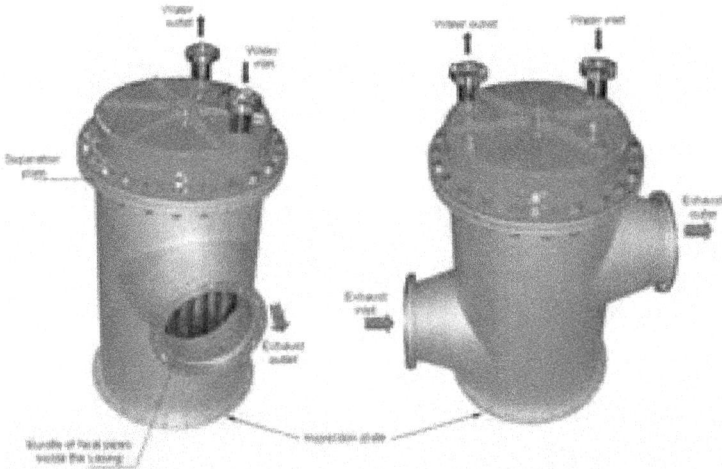

Figure 3.9: Commercial Heat Pipe Heat Exchanger (Econotherm, UK)

(usually, stainless steel to guard against rust), serves as the heat transfer surface. It is either rotated or kept stationary while hot and cold fluid alternately pass over it and is, thus, exposed, periodically, to the hot and cold streams. This enables it to cool the hot stream and heat up the cold stream in each cycle. The two common types of periodic flow heat exchangers are:

(a) The Rotary Type

The fluid flow here can be in either the axial or rotary direction with respect to the axis of rotation of the metal matrix. In schematic form these arrangements may be illustrated as shown in Figure 3.10.

Figure 3.10: Schematics of Axial and Radial Flow in Rotary Type Periodic Heat Exchangers

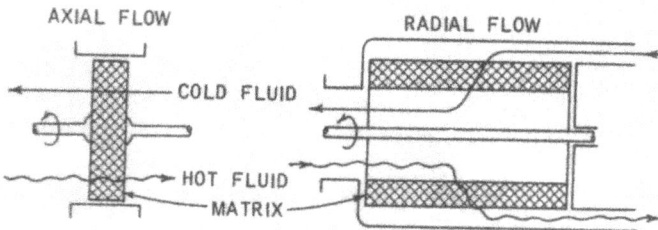

(b) The Valved Type

In this type, the matrix materials are fixed in space but hot and cold fluid are manipulated by means of valves to flow alternately through them to be cooled or heated on each passage. The periodic opening and closing of valves enables each of the two identical matrices to act alternately as hot and cold flow matrices. Schematically, the system may be represented as shown in Figure 3.11.

Figure 3.11: Schematic of Valved Type Periodic Heat Exchangers

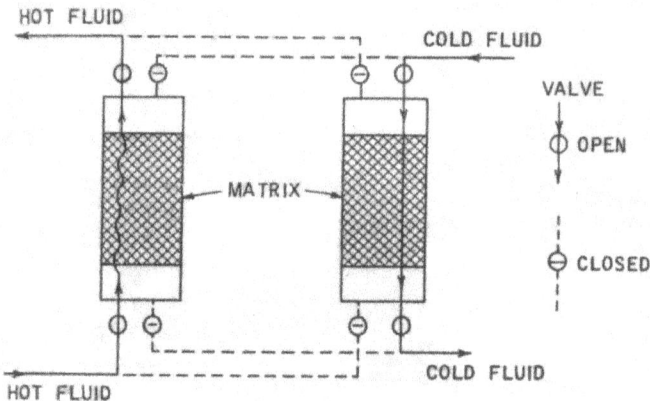

4. Direct Contact Heat Exchangers

In direct contact heat exchangers heat is transferred between hot and cold streams (phases) without a solid separating surface between them. Such phases may be

- Gas – liquid
- Immiscible liquid – liquid
- Solid-liquid or solid – gas

Most direct contact heat exchangers are used in gas-liquid operations where one of the streams is a gas and the other a liquid in the form of drops, films or sprays. Such operations occur in air conditioning, humidification, water cooling and condensing plants. Table 3.1 summarises the major configurations used in such systems.

Table 3.1: Configurations in Direct Contact Heat Exchangers (Wikipedia, 2010)

Phases	Continuous phase	Driving force	Change of phase	Examples
Gas – Liquid	Gas	Gravity	No	Spray columns, packed columns
			Yes	Cooling towers, falling droplet evaporators
		Forced	No	Spray coolers/quenchers
		Liquid flow	Yes	Spray condensers/evaporation, jet condensers
	Liquid	Gravity	No	Bubble columns, perforated tray columns
			Yes	Bubble column condensers
		Forced	No	Gas spargers
		Gas flow	Yes	Direct contact evaporators, submerged combustion

3.2: Heat Exchanger Arrangements and Configurations

Heat exchangers, especially, the direct type, are rarely used, in commercial practice, as simple or single units except in very simple processing situations. This is because the magnitude and complexity of heating and cooling duty in manufacturing or other commercial industry require more compact, efficient and hence more economic arrangements.

The first effort in this direction is to have more than one heat exchanger packaged as a single unit. This is the so called multi-pass heat exchangers, more commonly used in shell and tube heat exchanger configurations. The schematic for a 1-shell 2 tube-pass heat exchanger, for example, is shown in Figure 3.12.

Figure 3.12: Schematic of a 1-Shell 2 Tube-pass (1:2) Heat Exchanger

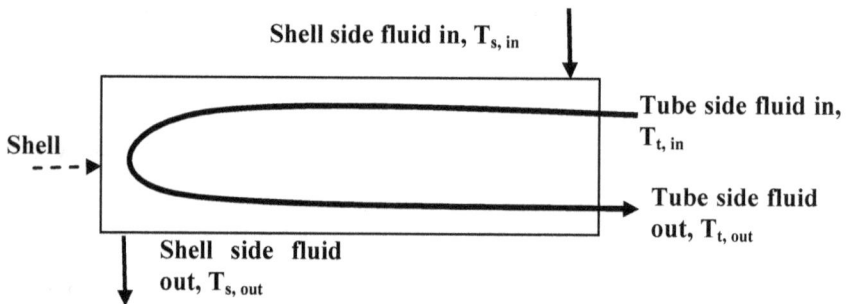

Shell side fluid in, $T_{s,\,in}$

Tube side fluid in, $T_{t,\,in}$

Shell

Tube side fluid out, $T_{t,\,out}$

Shell side fluid out, $T_{s,\,out}$

More complex multi-tube and multi-pass shell and tube arrangements require systematic nomenclature and standardization. These have been specified by the Tubular Exchangers Manufacturers Association (TEMA) and are illustrated in Figure 3.13.

The second practice is to have an array of heat exchangers, any of which may be single, multi shell or or multi-tube pass heat exchanger, arranged in such a way as to optimize heat exchange throughout the factory. Thus one heat exchanger may use the cooling fluid emanating from another heat exchanger while another may use the cooled liquid from one exchanger to heat up another fluid in a different heat exchanger. Such an arrangement is called a heat exchanger network.

Fluid compatibility with heat exchanger material is usually an issue. An example of a heat exchanger network is shown in Figure 3.14. The heat exchangers are labelled HE-1 to HE-6 in this illustration.

Figure 3.13: TEMA Classifications of Tubular Heat Exchangers

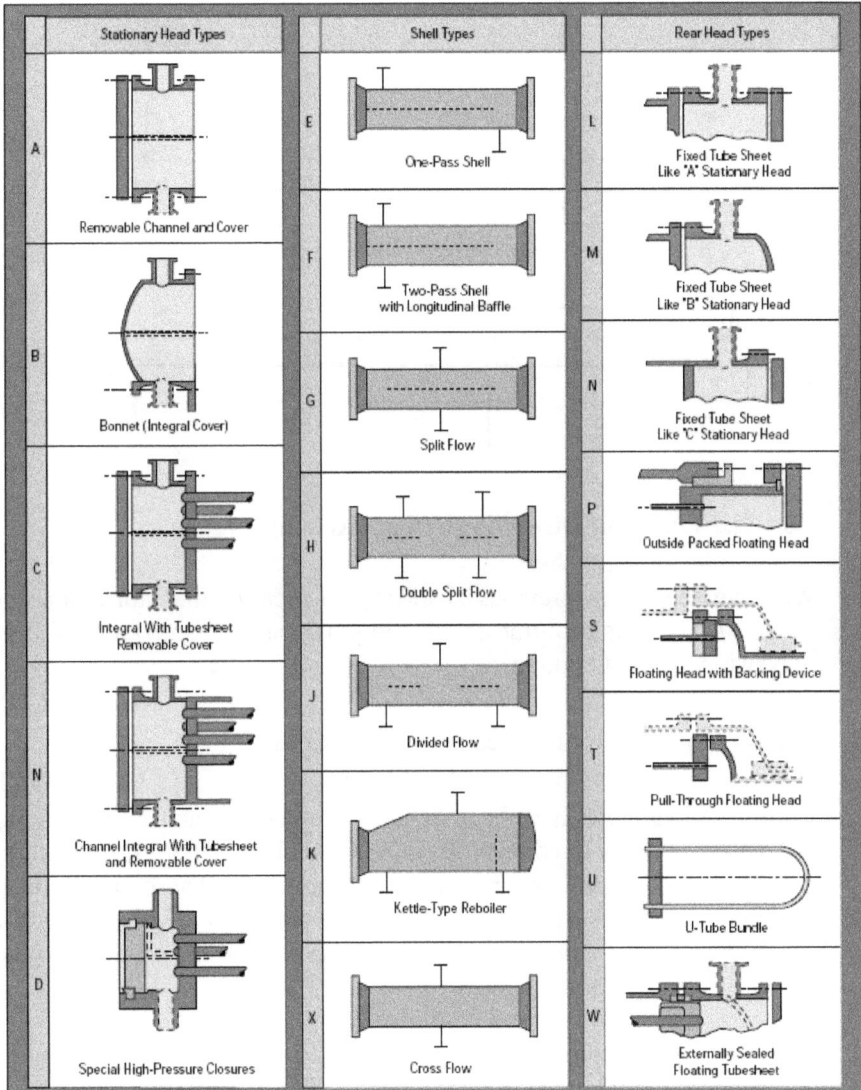

Stationary Head Types	Shell Types	Rear Head Types
A — Removable Channel and Cover	E — One-Pass Shell	L — Fixed Tube Sheet Like "A" Stationary Head
B — Bonnet (Integral Cover)	F — Two-Pass Shell with Longitudinal Baffle	M — Fixed Tube Sheet Like "B" Stationary Head
C — Integral With Tubesheet Removable Cover	G — Split Flow	N — Fixed Tube Sheet Like "C" Stationary Head
N — Channel Integral With Tubesheet and Removable Cover	H — Double Split Flow	P — Outside Packed Floating Head
D — Special High-Pressure Closures	J — Divided Flow	S — Floating Head with Backing Device
	K — Kettle-Type Reboiler	T — Pull-Through Floating Head
	X — Cross Flow	U — U-Tube Bundle
		W — Externally Sealed Floating Tubesheet

Source:

CHEMICAL ENGINEERING PROGRESS • FEBRUARY 1998

The shell and tube type heat exchanger is usually specified using a three letter code (from the TEMA specifications shown in Figure 3.13). Each

91

code describes a section of the heat exchanger such as the stationary head, the shell and the rear head type, eg AEL or DXW exchanger.

Figure 3.14: Schematic of a Six Heat Exchanger Network

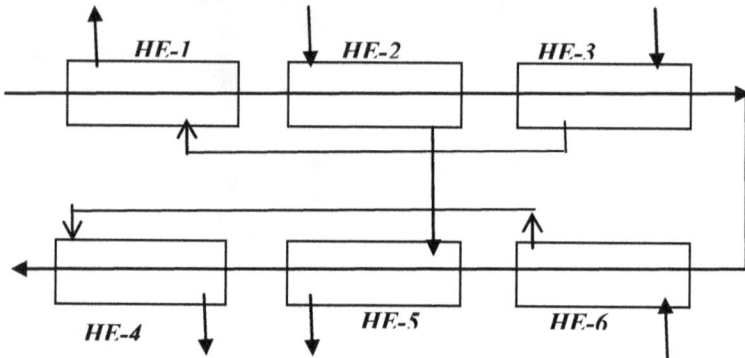

3.3: The Theory of Direct Type Heat Exchangers

We have seen from previous chapters and sections that, for simple or single surfaces, the heat transferred across the surface is given by a form of Newton's law of cooling

$$Q = q.A = h\Delta T_1.A \qquad (3.1)$$

and, consequently, for more complex surfaces by

$$Q = U.A.\Delta T_2 \qquad (3.2)$$

where these variables have been defined in previous chapters. In order to use these equations in commercial and practical applications, we need to develop some general theory which enables us to determine the heat transferred, Q, the surface area, A, and the temperature values, ΔT, for any system we may be considering and the relationship of these to the values of h and U for the system.

Consider, for example, the following situation in which a fluid is heated or cooled as it flows inside a closed conduit such as a circular pipe.

Figure 3.14: Schematic of Heat Transfer in Closed Conduits

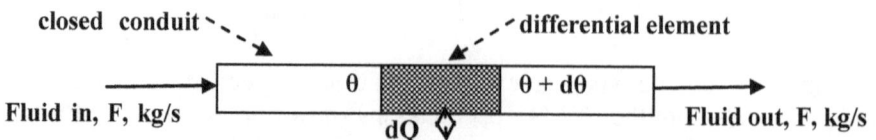

In the small differential element shown, over a small increment of time, *dt*

$$Heat\,added = d\,Q.d\,t = F.d\,t.Cp.d\,\theta \qquad (3.3)$$

where

$$
\begin{aligned}
t &= \text{time} \\
\theta &= \text{temperature at entrance of fluid into the element} \\
Cp_i &= \text{specific heat capacity over the element} \\
dQ &= \text{heat transfer rate into the element} \\
F &= \text{fluid flow rate}
\end{aligned}
$$

This gives us that

$$Q = F \int_{\theta_{out}}^{\theta_{in}} Cp_i \, d\theta \qquad (3.4)$$

The heat capacity, Cp, is, usually, expressed as a function of temperature. A mean heat capacity can, however, be found which is valid over the temperature range in which heat exchange occurs. Using this mean heat capacity, for simplicity and convenience, if it gives the level of accuracy that can be tolerated by the problem at hand, we can integrate equation (3.4) to obtain

$$Q = F\,\overline{Cp}\,(\theta_{out} - \theta_{in}) = F(H_{out} - H_{in}) \qquad (3.5)$$

where H is the specific enthalpy and $\overline{Cp}$ is the mean heat capacity in the temperature range. When we plot Q versus θ of equation (3.5), since F and Cp are constant, we get a straight line with slope = F. Cp

Fig. 3.15: Heating/Cooling Curve in a Heat Exchanger (Constant Heat Capacity)

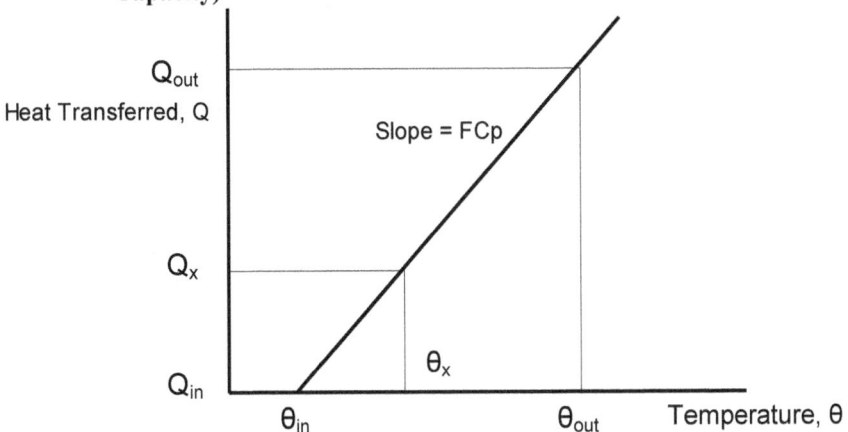

In practice, however, Cp is not always constant over the range of

temperatures in which the heat exchanger is operated and a mean value sometimes does not lead to an accurate estimate of the heat transferred. In such cases, the Q versus θ curve is not a straight line and the more general picture is that presented below.

Figure 3.16: Heating/Cooling Curve in a Heat Exchanger (Varying Heat Capacity)

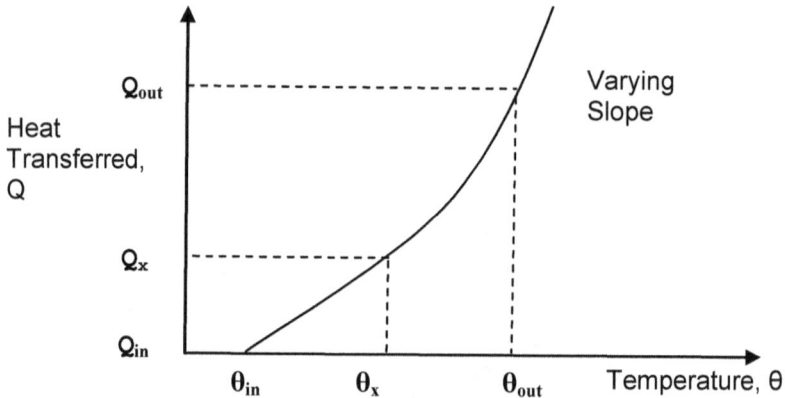

We can, thus, represent the Q versus θ curve of a liquid, such as water, in which phase change occurs in the exchanger, in heating or in cooling, as

Figure 3.17: Heating/Cooling Curve in Heat Exchangers (Phase Change)

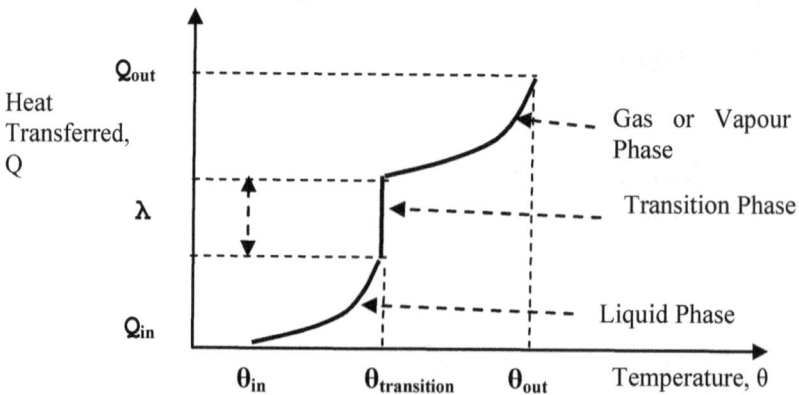

3.3.1: Exchange Diagrams

The above graphical relationships between Q and θ suggest that greater

94

insights may be gained by exploring this scheme further. This is the rationale for these diagrams which have come to be called exchange diagrams. Let us start with the simplest possible direct type heat exchanger, that is, a concentric pipe, also called a double pipe, heat exchanger. If the flow of the hot and cold fluids through this exchanger is co-current, the fluid flow directions and the exchange diagram are illustrated in Figure 3.18 below. We label the fluid entrance as section 1 and exit section as section 2, the heated fluid as fluid 1 and the cooled fluid as fluid 2. Then θ_{11}, θ_{21} become the entrance temperatures of the two fluids and θ_{12}, θ_{22} the exit temperatures. At any distance within the exchanger, we let the fluid temperatures be θ_1 and θ_2 respectively.

Figure 3.18: Fluid and Section Labels for Co-current Heat Exchangers

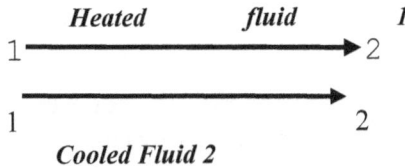

The temperature profile along the length of the heat exchanger will be as shown in Figure 3.19. The temperature difference, at any distance, x, along the exchanger, will be $\theta_1 - \theta_2$. Let us label this temperature difference, φ. This temperature difference, φ, will decrease along the length of the exchanger approaching zero asymptotically.

Figure 3.19: Temperature Profile in a Co-current Heat Exchanger

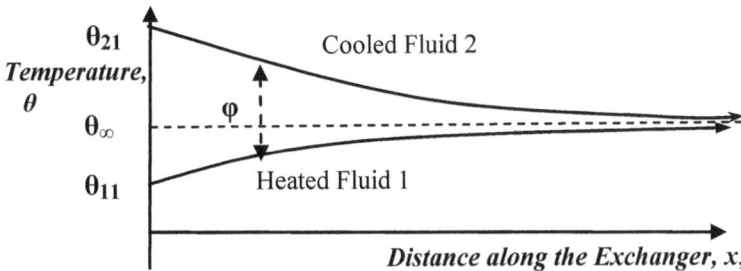

The exchange diagram, Fig 3.20, is even more interesting. It shows that there is a temperature limit, θ_∞, at $\varphi = 0$. It shows, also, that the maximum possible heat transferable in this exchanger cannot be greater than Q_∞.

Figure 3.20: Exchange Diagram for a Co-current Heat Exchanger

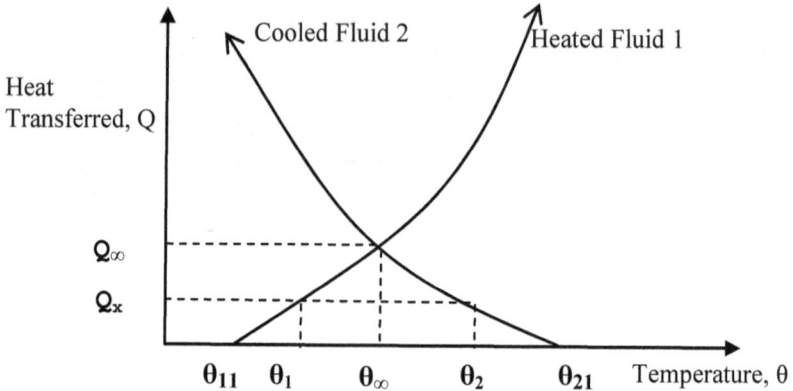

In counter-current flow, the fluid flow directions, the temperature profile along the heat exchanger and the exchange diagram are illustrated below.

Figure 3.21: Fluid and Section Labels for Countercurrent Heat Exchangers

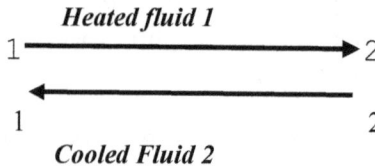

Figure 3.22: Temperature Profile in a Countercurrent Heat Exchanger

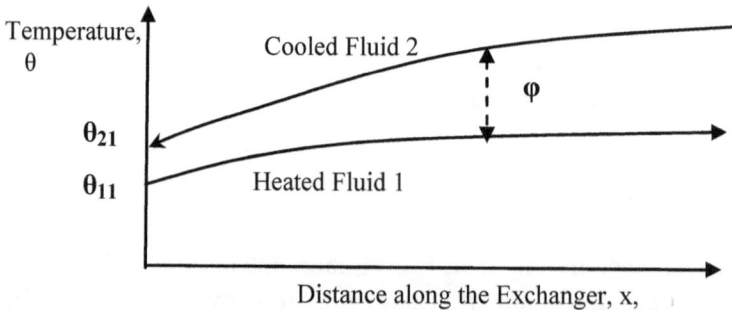

It would appear, from Figure 3.22, that, within the range in which normal heat exchangers operate, there appears to be no limit to Q.

Figure 3.23: Exchange Diagram for a Counter-current Heat Exchanger

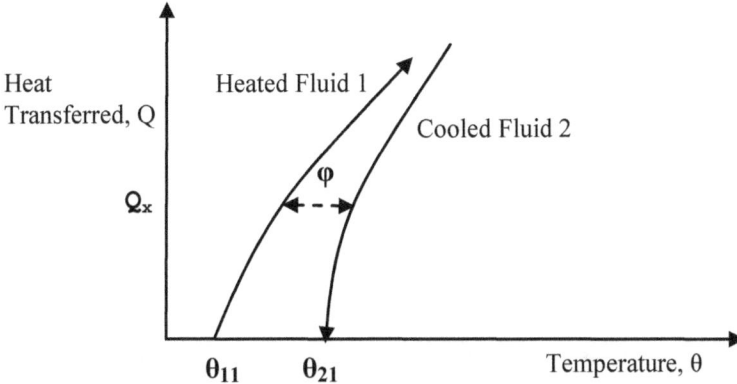

The exchange diagram shows, however, that as Q increases, the outlet temperature of the heated fluid approaches the inlet temperature of the cooled fluid. This implies that there will, indeed, be a limit to the heat transfer, Q_{limit}, which will occur at the hot or cold end of the heat exchanger depending on whether $F_H Cp_H$ is greater or less than $F_C Cp_C$. The subscript H represents the hot or cooled fluid while C represents the cold or heated fluid, at inlet.

3.3.2: The Log Mean Temperature Difference

If the heat capacity is not constant throughout the heat exchanger and for both fluids, Figures 3.19 and 3.22 tell us that the temperature difference, φ, will vary along the length of the heat exchanger. Of course, we can always attempt to estimate this temperature difference, $\theta_1 - \theta_2$, for every infinitesimal section of the heat exchanger, calculate the corresponding heat transfer coefficient and surface area and sum all the heat transferred at each section to obtain the overall heat transferred, Q.

There are, in fact, situations in which this is the only procedure that will be possible. Generally, however, having defined Q, in terms of an overall heat transfer coefficient, U, as in equation (2.16), we would need to define a temperature difference, ΔT, which can be used in this equation to obtain the correct estimate of Q. The first temptation is to use the various mathematical means, the so-called Pythagorean means (arithmetic, geometric and the harmonic) or the logarithmic, power or root means, each posing its own peculiar problems of complexity.

We can, however, consider what actually happens to the temperature difference along the exchanger. Consider a countercurrent two fluid heat exchanger as a, typically, general case.

Figure 3.24: Schematic Diagram for Temperature Differences in a Counter-current Heat Exchanger

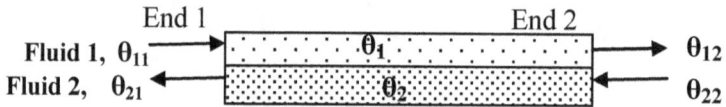

Let the total surface area of the heat exchanger be S. Over this surface area S

$$Q_T = F_1\, Cp_1\, (\theta_{12} - \theta_{11}) = F_2\, Cp_2\, (\theta_{22} - \theta_{21}) \qquad (3.6)$$

where Q_T is the total heat transferred. At any distance, x, within the heat exchanger, let the fluid temperatures be θ_1 and θ_2. Then, for constant heat capacity at this distance

$$Q_x = F_1\, Cp_1\, (\theta_1 - \theta_{11}) = F_2\, Cp_2\, (\theta_2 - \theta_{21}) \qquad (3.7)$$

We can get the values of θ_1 and θ_2 from equations (3.7) as

$$\theta_1 = \theta_{11} + \frac{Q_x}{F_1\, Cp_1} \qquad (3.8)$$

$$\theta_2 = \theta_{21} + \frac{Q_x}{F_2\, Cp_2} \qquad (3.9)$$

and the temperature difference, $\varphi = \theta_2 - \theta_1$, at any cross section of the exchanger, as

$$\varphi = \theta_2 - \theta_1 = (\theta_{21} - \theta_{11}) - Q_x\left(\frac{1}{F_1\, Cp_1} - \frac{1}{F_2\, Cp_2}\right) \qquad (3.10)$$

If we let

$$m = \frac{1}{F_1\, Cp_1} - \frac{1}{F_2\, Cp_2} \qquad (3.11)$$

then

$$\varphi = \varphi_1 - m Q_x \qquad (3.12)$$

Recall that

$$Q_x = U_x\, S_x\, \varphi \quad \text{so that} \quad dQ = U_x\, \varphi\, dS \qquad (3.13)$$

For constant U

$$S = \frac{1}{U} \int_0^{Q_T} \frac{dQ}{\varphi} \qquad (3.13a)$$

Substituting for φ, from equation (3.12), into equation (3.13a) and integrating, we get that, for any Q,

$$S = \frac{1}{U} \int_0^Q \frac{dQ}{\varphi} = \frac{1}{U} \int_0^Q \frac{dQ}{\varphi_1 - mQ} = \frac{1}{mU} \ln\left(\frac{\varphi_1}{\varphi - mQ}\right) \qquad (3.13b)$$

Equation (3.13b) can, also, be stated as

$$S = \frac{1}{mU} \ln\left(\frac{\varphi_1}{\varphi_1 - mQ}\right) = \frac{1}{mU} \ln\left(\frac{\varphi_1}{\varphi}\right) \quad \text{or as} \quad \varphi = \varphi_1 e^{-mUS} \qquad (3.13c)$$

If an overall mean temperature difference, φ_M, is defined such that

$$Q_T = U_T S_T \varphi_M \quad \text{then} \quad \varphi_M = \frac{Q_T}{U_T S_T} \qquad (3.14)$$

Substituting equation (3.14) into (3.13c)

$$S_T = \frac{Q_T}{U_T \varphi_M} = \frac{1}{mU_T} \ln\left(\frac{\varphi_1}{\varphi_T}\right) \quad \text{or} \quad \varphi_M = \frac{mQ_T}{\ln\left(\frac{\varphi_1}{\varphi_T}\right)} \qquad (3.14a)$$

But from equation (3.12)

$$m = \frac{\varphi_1 - \varphi_T}{Q_T} \qquad (3.12a)$$

Hence, from equations (3.12a) and (3.14a)

$$\varphi_M = \frac{mQ_T}{\ln\left(\frac{\varphi_1}{\varphi_T}\right)} = \frac{\varphi_1 - \varphi_T}{\ln\left(\frac{\varphi_1}{\varphi_T}\right)} \qquad (3.15)$$

φ_M is seen to be a logarithmic mean temperature difference. The use of this temperature difference is valid only if U is constant and φ varies linearly with Q. It is not applicable to condensers, reboilers and to situations where the latent heat, associated with phase change, is involved.

Illustrative Example

A hot fluid enters a direct type two fluid heat exchanger at 200 °C and is cooled to 150 °C. In the process, the cold fluid, which entered at 80 °C, is heated to 120 °C. Calculate the log mean temperature difference for

(a) Co-current flow   (b) Counter current flow.

99

Answer

(a) Underline Cocurrent Flow

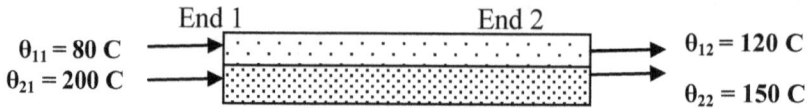

$$\varphi_M = \frac{\varphi_1 - \varphi_T}{\ln\left(\dfrac{\varphi_1}{\varphi_T}\right)} = \frac{(\theta_{21}-\theta_{11})-(\theta_{22}-\theta_{12})}{\ln\left(\dfrac{\theta_{21}-\theta_{11}}{\theta_{22}-\theta_{12}}\right)} = \frac{(200-80)-(150-120)}{\ln\left(\dfrac{200-80}{150-120}\right)} = 64.92C$$

(b) Countercurrent Flow

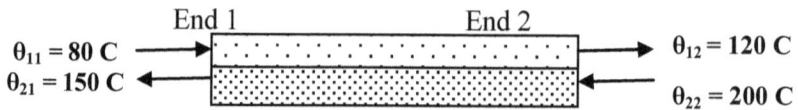

$$\varphi_M = \frac{\varphi_1 - \varphi_T}{\ln\left(\dfrac{\varphi_1}{\varphi_T}\right)} = \frac{(\theta_{21}-\theta_{11})-(\theta_{22}-\theta_{12})}{\ln\left(\dfrac{\theta_{21}-\theta_{11}}{\theta_{22}-\theta_{12}}\right)} = \frac{(150-80)-(200-120)}{\ln\left(\dfrac{150-80}{200-120}\right)} = 74.89C$$

It can be seen that counter-current flow offers greater log mean temperature difference, hence, higher heat transfer than co-current flow.

3.3.3: Heat Exchanger Surface Area

Equation (3.14) shows that the surface area, based on an overall heat transfer coefficient, U, and a logarithmic temperature difference, φ_M, is given by

$$S_T = \frac{Q_T}{U_T\,\varphi_M} = \frac{Q_T}{U_T\left(\varphi_1 - \varphi_T\right)}\ln\left(\frac{\varphi_1}{\varphi_T}\right) \qquad (3.16)$$

3.3.4: Heat Exchanger Effectiveness and Number of Transfer Units

There is often the need to determine how energy efficient a heat exchanger is. This enables a choice to be made between possible configurations in a direct type heat exchanger. Thermal efficiency is, particularly, important in the design of compact heat exchangers used in domestic and household appliances as well as in high technology electronics and control applications. Thermal efficiency of heat exchangers is often defined in terms of a) effectiveness, ε, and b) the number of transfer units, NTU.

The heat transfer effectiveness of a heat exchanger is defined as the ratio of the actual heat transferred to the maximum possible heat transferrable in that exchanger. That is

$$Effectiveness, \varepsilon, = \frac{Q_{actual}}{Q_{max}}$$
$$= \frac{C_h\left(\theta_{h,in} - \theta_{h,out}\right)}{C_{min}\left(\theta_{h,in} - \theta_{c,in}\right)} = \frac{C_c\left(\theta_{c,out} - \theta_{c,in}\right)}{C_{min}\left(\theta_{h,in} - \theta_{c,in}\right)} \quad (3.17)$$

The subscript h denotes the hot or cooled stream while subscript c denotes the cold or heated stream. Fluid heat capacity, $C = F \cdot Cp$ while C_{min} is the smaller of C_h and C_c. Usually, but not always, $C_h < C_c$. Recall that

$$dQ = F.Cp.d\theta = U.\varphi.dS \quad that is \quad \frac{d\theta}{\varphi} = \frac{U.dS}{F.Cp} \quad (3.18)$$

The number of transfer units, NTU, is, consequently, defined as

$$NTU = \int_{\theta_{in}}^{\theta_{out}} \frac{d\theta}{\varphi} = \int_0^S \frac{U.dS}{F.Cp} = \frac{U.S}{C_{min}} \quad (3.19)$$

3.3.4.1: Effectiveness – NTU Relations

A look at the formulae for effectiveness and the NTU shows that both can be used to define a heat exchanger in the usual terms of temperature changes, fluid flowrates and heat exchanger surface in addition to indicating the thermal efficiency of such an exchanger. It makes sense, therefore, to find a relationship between these two terms. Let us start with the simplest configuration, the double pipe heat exchanger in co-current and in counter-current flow.

Effectiveness – NTU Relation for a Co-current Heat Exchanger

Consider the following schematic diagram of a co-current heat exchanger in which we consider the exchange of heat between two fluids at any point in the exchanger. We shall assume that heat is being transferred from fluid 2 to fluid 1 and that $C_1 < C_2$.

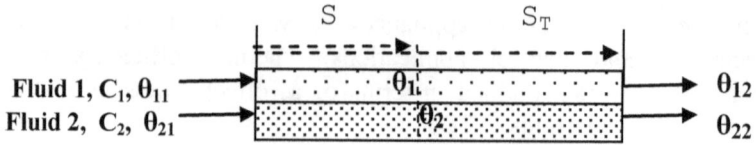

A heat energy balance up to surface area, S, gives

$$Q = C_1 \left(\theta_1 - \theta_{11}\right) = C_2 \left(\theta_{21} - \theta_2\right) \qquad (3.20)$$

That is $\qquad \theta_2 = \theta_{21} - \dfrac{Q}{C_2} \quad and \quad \theta_1 = \theta_{11} + \dfrac{Q}{C_1}$

or

$$\varphi = \theta_2 - \theta_1 = \theta_{21} - \frac{Q}{C_2} - \theta_{11} - \frac{Q}{C_1} = \theta_{21} - \theta_{11} - Q\left(\frac{C_1 + C_2}{C_1 C_2}\right)$$

$$= \theta_{21} - \theta_{11} - Q\left(\frac{\mu+1}{C_1}\right) = \varphi_1 - Q\left(\frac{\mu+1}{C_1}\right) \qquad (3.21)$$

where $\mu = C_1/C_2$, called the capacity ratio and $\varphi_1 = \theta_{21} - \theta_{11}$. We can substitute equation (3.21) into equation (3.13 to get

$$S_T = \frac{1}{U}\int_0^{Q_T} \frac{dQ}{\varphi} = \frac{1}{U}\int_0^{Q_T} \frac{dQ}{\left(\varphi_1 - \dfrac{Q(\mu+1)}{C_1}\right)}$$

$$= \frac{C_1}{U(\mu+1)} \ln \frac{\varphi_1}{\varphi_1 - \dfrac{Q_T(\mu+1)}{C_1}} \qquad (3.22)$$

A heat energy balance up to surface area, S_T, gives

$$Q_T = C_1 \left(\theta_{12} - \theta_{11}\right) = C_2 \left(\theta_{21} - \theta_{22}\right) \qquad (3.23)$$

That is $\qquad \theta_{22} = \theta_{21} - \dfrac{Q_T}{C_2} \quad and \quad \theta_{12} = \theta_{11} + \dfrac{Q_T}{C_1}$

$$giving \ \varphi_2 = \theta_{22} - \theta_{12} = \theta_{21} - \frac{Q_T}{C_2} - \theta_{11} - \frac{Q_T}{C_1} = \theta_{21} - \theta_{11} - Q_T\left(\frac{C_1 + C_2}{C_1 C_2}\right)$$

$$= \varphi_1 - \frac{Q_T(\mu+1)}{C_1} \quad or \quad \frac{Q_T(\mu+1)}{C_1} = \varphi_1 - \varphi_2 \qquad (3.24)$$

Substituting equation (3.24) into equation (3.22) we get that

$$S_T = \frac{C_1}{U(\mu+1)} \ln \frac{\varphi_1}{\varphi_2} \qquad (3.25)$$

But NTU is defined, from equation (3.19), as

102

$$NTU = \int_{\theta_{in}}^{\theta_{out}} \frac{d\theta}{\varphi} = \int_0^S \frac{U.dS}{F.Cp} = \frac{U.S}{C_{min}} \qquad \text{from (3.19)}$$

Hence, from equations (3.25) and (3.19)

$$\frac{US}{C_1} = NTU = \frac{1}{(\mu+1)} \ln\frac{\varphi_1}{\varphi_2} \quad i.e \quad \varphi_2 = \varphi_1\, e^{-NTU\,(\mu+1)} \quad (3.26)$$

That is

$$\theta_{22} - \theta_{12} = (\theta_{21} - \theta_{11})e^{-NTU\,(\mu+1)} \qquad (3.26a)$$

From the definition of heat exchanger effectiveness, from equation (3.17)

$$\varepsilon = \frac{\mu(\theta_{21} - \theta_{22})}{(\theta_{21} - \theta_{11})} = \frac{(\theta_{12} - \theta_{11})}{(\theta_{21} - \theta_{11})} \qquad (3.27)$$

Now

$$\theta_{21} - \theta_{22} = \theta_{21} - \theta_{11} + \theta_{11} - \theta_{22}$$
$$= (\theta_{21} - \theta_{11}) - \theta_{22} + \theta_{11} + \theta_{12} - \theta_{12}$$
$$= (\theta_{21} - \theta_{11}) - (\theta_{22} - \theta_{12}) - (\theta_{12} - \theta_{11})$$
$$= (\theta_{21} - \theta_{11}) - (\theta_{21} - \theta_{11})e^{-NTU\,(\mu+1)} - \mu(\theta_{21} - \theta_{22}) \quad (3.28)$$

That is

$$(1+\mu)(\theta_{21} - \theta_{22}) = (\theta_{21} - \theta_{11})\left(1 - e^{-NTU\,(\mu+1)}\right)$$

$$or \quad \frac{(\theta_{21} - \theta_{22})}{(\theta_{21} - \theta_{11})} = \frac{\varepsilon}{\mu} = \frac{\left(1 - e^{-NTU\,(\mu+1)}\right)}{(1+\mu)} \qquad (3.28a)$$

Finally

$$\varepsilon = \frac{\mu\left(1 - e^{-NTU\,(\mu+1)}\right)}{(1+\mu)} \qquad (3.29)$$

Effectiveness – NTU Relation for a Countercurrent Heat Exchanger

Similarly, for a countercurrent heat exchanger, we continue to assume that heat is being transferred from fluid 2 to fluid 1 and that $C_1 < C_2$.

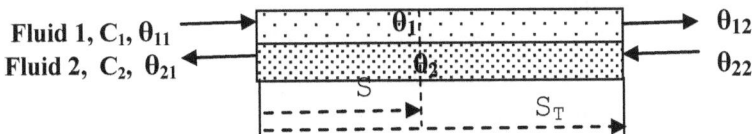

A heat energy balance up to surface area, S, gives

$$Q = C_1 (\theta_1 - \theta_{11}) = C_2 (\theta_2 - \theta_{21}) \qquad (3.30)$$

That is

$$\theta_2 = \theta_{21} + \frac{Q}{C_2} \quad and \quad \theta_1 = \theta_{11} + \frac{Q}{C_1}$$

$$giving \; \varphi = \theta_2 - \theta_1 = \theta_{21} + \frac{Q}{C_2} - \theta_{11} - \frac{Q}{C_1} = \theta_{21} - \theta_{11} - Q\left(\frac{C_2 - C_1}{C_1 C_2}\right)$$

$$= \theta_{21} - \theta_{11} - Q\left(\frac{1-\mu}{C_1}\right) = \varphi_1 - Q\left(\frac{1-\mu}{C_1}\right) \tag{3.31}$$

where $\mu = C_1/C_2$, called the capacity ratio and $\varphi_1 = \theta_{21} - \theta_{11}$. We can substitute equation (3.31) into equation (3.13 to get

$$S_T = \frac{1}{U}\int_0^{Q_T} \frac{dQ}{\varphi} = \frac{1}{U}\int_0^{Q_T} \frac{dQ}{\left(\varphi_1 - \frac{Q(1-\mu)}{C_1}\right)}$$

$$= \frac{C_1}{U(1-\mu)} \ln \frac{\varphi_1}{\varphi_1 - \frac{Q_T(1-\mu)}{C_1}} \tag{3.32}$$

A heat energy balance up to surface area, S_T, gives

$$Q_T = C_1\left(\theta_{12} - \theta_{11}\right) = C_2\left(\theta_{22} - \theta_{21}\right) \tag{3.33}$$

That is

$$\theta_{22} = \theta_{21} + \frac{Q_T}{C_2} \quad and \quad \theta_{12} = \theta_{11} + \frac{Q_T}{C_1}$$

$$giving \; \varphi_2 = \theta_{22} - \theta_{12} = \theta_{21} + \frac{Q_T}{C_2} - \theta_{11} - \frac{Q_T}{C_1} = \theta_{21} - \theta_{11} - Q_T\left(\frac{C_2 - C_1}{C_1 C_2}\right)$$

$$= \varphi_1 - \frac{Q_T(1-\mu)}{C_1} \quad or \quad \frac{Q_T(1-\mu)}{C_1} = \varphi_1 - \varphi_2 \tag{3.34}$$

Substituting equation (3.34) into equation (3.32) we get that

$$S_T = \frac{C_1}{U(1-\mu)} \ln \frac{\varphi_1}{\varphi_2} \tag{3.35}$$

But NTU is defined as

$$NTU = \int_{\theta_{in}}^{\theta_{out}} \frac{d\theta}{\varphi} = \int_0^S \frac{U.dS}{F.Cp} = \frac{U.S}{C_{min}} \tag{3.19}$$

Hence, from equations (3.35) and (3.19)

$$\frac{U S}{C_1} = NTU = \frac{1}{(1-\mu)} \ln \frac{\varphi_1}{\varphi_2} \quad i.e \quad \varphi_2 = \varphi_1 e^{-NTU(1-\mu)} \tag{3.36}$$

That is

$$\theta_{22} - \theta_{12} = (\theta_{21} - \theta_{11})e^{-NTU(1-\mu)} \tag{3.37}$$

From the definition of heat exchanger effectiveness, from equation (3.17)

$$\varepsilon = \frac{\mu(\theta_{22} - \theta_{21})}{(\theta_{22} - \theta_{11})} = \frac{(\theta_{12} - \theta_{11})}{(\theta_{22} - \theta_{11})} \tag{3.38}$$

Now
$$\theta_{22} - \theta_{21} = \theta_{22} - \theta_{11} + \theta_{11} - \theta_{21}$$
$$= (\theta_{22} - \theta_{11}) - \theta_{21} + \theta_{11} + \theta_{12} - \theta_{12}$$
$$= (\theta_{22} - \theta_{11}) - \theta_{21} + \theta_{12} - \theta_{12} + \theta_{11} + \theta_{22} - \theta_{22}$$
$$= (\theta_{22} - \theta_{11}) - (\theta_{12} - \theta_{11}) - (\theta_{22} - \theta_{12}) + (\theta_{22} - \theta_{21}) \tag{3.39}$$

Thus, using equations (3.37) and (3.38),

$$\mu(\theta_{22} - \theta_{21}) = (\theta_{22} - \theta_{11})\left(1 - e^{-NTU\,(1-\mu)}\right)$$

$$or \quad \frac{(\theta_{22} - \theta_{21})}{(\theta_{22} - \theta_{11})} = \frac{\varepsilon}{\mu} = \frac{\left(1 - e^{-NTU\,(1-\mu)}\right)}{\mu} \tag{3.40}$$

Finally,

$$\varepsilon = 1 - e^{-NTU\,(1-\mu)} \tag{3.41}$$

Similar relationships can be developed for other configurations of heat exchangers. Kays & London, (1958), for example have listed a number of these and some are shown in Tables 3.2 to 3.5.

Table 3.2:   NTU – Effectiveness Relations for Direct Type Heat Exchangers (Kays & London, 1958)

Cocurrent Flow	$\varepsilon = \dfrac{1 - e^{-NTU\,(1+\mu)}}{(1+\mu)}$
Countercurrent Flow	$\varepsilon = \dfrac{1 - e^{-NTU\,(1-\mu)}}{1 - \mu e^{-NTU\,(1-\mu)}}$
Cross flow (one fluid mixed, the other unmixed)	$\varepsilon = \dfrac{1}{\mu}\left(1 - e^{-p\mu}\right) \quad where \quad p = 1 - e^{-NTU}$
Cross flow (both fluids mixed)	$\varepsilon = \dfrac{NTU}{\dfrac{NTU}{1 - e^{-NTU}} + \dfrac{\mu\,NTU}{1 - e^{-\mu NTU}} - 1}$
Multipass overall counterflow (fluids mixed between passes)	$\varepsilon = \dfrac{\left(\dfrac{1 - \mu\varepsilon_{pass}}{1 - \varepsilon_{pass}}\right)^{n} - 1}{\left(\dfrac{1 - \mu\varepsilon_{pass}}{1 - \varepsilon_{pass}}\right)^{n} - \mu}$

Table 3.3: NTU – Effectiveness Relations for Indirect Transfer Liquid Coupled Heat Exchangers (Heat Capacities of the Hot and Cold Fluids are Equal) (Kays & London, 1958)

$C_c = C_h = C > C_L$	$\varepsilon = \dfrac{C_L/C}{\dfrac{1}{\varepsilon_h} + \dfrac{1}{\varepsilon_c} - 1}$
$C_c = C_h = C < C_L$	$\varepsilon = \dfrac{1}{\dfrac{1}{\varepsilon_c} + \dfrac{1}{\varepsilon_h} + \dfrac{C}{C_L}}$
$C_c = C_h = C_L = C$	$\varepsilon = \dfrac{1}{\dfrac{1}{\varepsilon_c} + \dfrac{1}{\varepsilon_h} - 1}$

Table 3.4: NTU – Effectiveness Relations for Indirect Transfer Liquid Coupled Heat Exchangers (Heat Capacities of the Hot and Cold Fluids are not Equal) (Kays & London, 1958)

$C_L > C_c > C_h$	$\varepsilon = \dfrac{1}{\dfrac{1}{\varepsilon_h} + \dfrac{C_h/C_c}{\varepsilon_c} + \dfrac{C_h}{C_L}}$
$C_L > C_h > C_c$	$\varepsilon = \dfrac{1}{\dfrac{1}{\varepsilon_c} + \dfrac{C_c/C_h}{\varepsilon_h} + \dfrac{C_c}{C_L}}$
$C_c > C_h > C_L$	$\varepsilon = \dfrac{1}{\dfrac{C_h}{C_L}\left(\dfrac{1}{\varepsilon_c} + \dfrac{1}{\varepsilon_h} - 1\right)}$
$C_h > C_c > C_L$	$\varepsilon = \dfrac{1}{\dfrac{C_c}{C_L}\left(\dfrac{1}{\varepsilon_c} + \dfrac{1}{\varepsilon_h} - 1\right)}$
$C_c > C_L > C_h$	$\varepsilon = \dfrac{1}{\dfrac{1}{\varepsilon_h} + \dfrac{C_h}{C_L}\left(\dfrac{1}{\varepsilon_c} - 1\right)}$
$C_h > C_L > C_c$	$\varepsilon = \dfrac{1}{\dfrac{1}{\varepsilon_c} + \dfrac{C_h}{C_L}\left(\dfrac{1}{\varepsilon_h} - 1\right)}$

Table 3.5: NTU - Effectiveness Relations in Periodic Flow Heat Exchangers (Kays & London, 1958)

Effectivesness	$\varepsilon = f\left(NTU, \mu_{fluids}, \mu_{matrix}\right)$
Number of Transfer Units	$NTU = \dfrac{1}{C_{min}} \left[\dfrac{1}{\left(\dfrac{1}{hS}\right)_c + \left(\dfrac{1}{hS}\right)_h} \right]$
Matrix capacity rate, C_R	$C_R = rpm \, x \, mass \, of \, matrix \, x \, sp.ht \, of \, matrix / 60$
Conductance ratio, $h*$	$h* = \dfrac{(hS) \, on \, C_{min} \, side}{(hS) \, on \, C_{max} \, side}$
Effectiveness at matrix speed, ε_R	$\varepsilon_R = \varepsilon_{counterflow \, direct \, type} \left[1 - \dfrac{1}{9\left(\dfrac{C_R}{C_{min}}\right)^{1.93}} \right]$

3.3.5: Pressure Drop in Heat Exchangers

Another major variable, in heat exchanger design or operation, is the pressure drop through the exchanger. Pressure drop affects the cost of operating the heat exchanger. If the pressure drop is too high, the cost of pumping fluids through it will be high. Thus, a very complex surface arrangement which gives very a high surface area per unit volume, or high effectiveness for a few NTU, may be associated with a very high pumping cost due to the high pressures required to pump either or both of the process or utility fluids through it.

Generally, for high density fluids, such as liquids, the heat transfer rate is much higher than the pressure drop. In these cases, pressure drop considerations are, usually, limited to specifying a maximum allowable pressure drop based on other cost factors in the plant or system. For low density fluids, such as gases, it is possible to spend as much mechanical energy overcoming friction as the heat energy being transferred.

With these considerations in mind, we can identify the main components of pressure drop in heat exchangers as

107

i).   Fluid friction arising from the flow of fluids through the exchanger. This friction can arise from flow inside or outside the tubes, as is the case with the so called shell and tube heat exchangers

ii).   Friction due to sudden expansion, contraction or reversal in flow direction (the so called entrance, exit, contraction and expansion losses)

iii).   Losses due to changes in kinetic and potential energy. Generally, these losses are negligible, especially for single units, since there is not much change in kinetic or potential energy along the length of standard sizes of heat exchangers.

The most significant causes of pressure drop in heat exchangers, however, are, first, fluid friction arising from the flow of fluids inside or outside the conduits, followed by the pressure drop due to entrance, exit and expansion losses. Because shell and tube heat exchangers constitute the work horses of manufacturing industry, and hence contribute to greater costs in pressure losses, greater attention has been paid to estimating their pressure drop characteristics than, perhaps, to other types of heat exchangers. Most of the treatment of pressure drop in heat exchangers here, therefore, will be based on those for shell and tube heat exchangers.

3.3.5.1: Pressure Drop inside Heat Exchanger Tubes

The most general expression for the pressure drop inside tubes is that which takes into account all the flow effects. Such an expression was given by Kays & London (1958) as

$$Entrance\,effects \quad Flow\,acceleration \quad Tube\,losses$$
$$\downarrow \qquad\qquad \downarrow \qquad\qquad \downarrow$$

$$\Delta P = u_1 \frac{G^2}{2}\left[\left(K_{entrance}+1-\sigma^2\right)+2\left(\frac{u_2}{u_1}-1\right)+f\left(\frac{A}{A_t}\right)\left(\frac{u_m}{u_1}\right)\right]$$

$$-\frac{u_2 G^2}{2}\left(1-\sigma^2-K_{exit}\right) \qquad\qquad (3.42)$$

$$\uparrow$$

$$Exit\,effects$$

where G is a mass rate of flow, u is the velocity, A the flow area, f, a friction factor, K a constant and σ is the ratio of free flow area to the frontal area of one side of the heat exchanger. The subscripts 1 and 2 denote the sections of the heat exchanger of interest, m, a mean value, and t, the continuous portion of the exchanger. How each component of the above equation is evaluated was given in some detail by the authors in the original text. A more explicit version of the above equation is, if header losses are neglected,

$$\Delta P = N_p\, \rho_i\, u_i^2 \left[4\frac{L}{D_i}\left(\frac{R}{\rho u^2}\right)_i + \frac{(K_{entrance} + K_{exit})}{2}\right] \tag{3.43}$$

where N_p is the number of tube passes, L is the length of tubes and D_i is the tube inside diameter. Sinnot (1983) relates this equation to the j-factor using the $R/\rho u^2$ which is a friction factor and obtains

$$\Delta P = N_p \left[8 j_f \cdot \frac{L}{D_i}\left(\frac{\mu_i}{\mu_w}\right)_i^{-m} \cdot \frac{\rho_i\, u_i^2}{2}\right] \tag{3.44}$$

where m = 0.25 for laminar flow, i.e. Re < 2100
        = 0.14 for turbulent flow, i.e. Re > 2100
μ_i = viscosity estimated at tube fluid temperature
μ_w = viscosity estimated at tube wall temperature
j_f is estimated from graphs of j_f versus Re such as those shown in Figures 3.25 and 3.26. Equation (3.44) is valid, also, in non-isothermal flow.

Peters & Timmerhaus (1985) give the expression

$$\Delta P = \frac{2 f_i\, N_p\, B_i\, L\, G^2}{\rho_i\, D_i\, \phi_i} \tag{3.45}$$

where f_i is the Fanning friction factor for isothermal flow based on the arithmetic mean temperature of the fluid. ϕ_i is the correction factor for non-isothermal flow and is equal $1.1\left(\frac{\mu_i}{\mu_w}\right)^{-0.25}$ for laminar flow (Re < 2100) and equal to $1.02\left(\frac{\mu_i}{\mu_w}\right)^{-0.14}$ for turbulent flow (Re > 2100).

For smooth pipes and turbulent flow

$$B_i = 1 + \frac{0.51 K_t\, N_p\, \Delta T_S \left(\frac{\mu_i}{\mu_w}\right)^{0.28}}{(T_2 - T_1)_i \left(\frac{Cp\,\mu}{k}\right)_i^{\frac{2}{3}}} \tag{3.46}$$

where

$$K_t = \left(1 - \frac{S_i}{S_{hd}}\right)^2 + K_C + 0.5\left(\frac{N_p - 1}{N_p}\right) \qquad (3.46a)$$

$$K_t = 0.4\left(1.25 - \frac{S_i}{S_{hd}}\right) \quad if \quad \frac{S_i}{S_{hd}} < 0.715 \qquad (3.46b)$$

$$= 0.75\left(1 - \frac{S_i}{S_{hd}}\right) \quad if \quad \frac{S_i}{S_{hd}} > 0.715 \qquad (3.46c)$$

S_i and S_{hd} are, respectively, the total inside and header cross-sectional area per pass. ΔT_S is the difference between the mean temperatures of the two streams exchanging heat energy. $T_2 - T_1$ is the temperature difference in the fluid flowing inside the tubes.

Fig. 3.25: Friction Factor vs Reynolds vs Reynolds Number (Flow inside Pipes) (Sinnot, 1983)

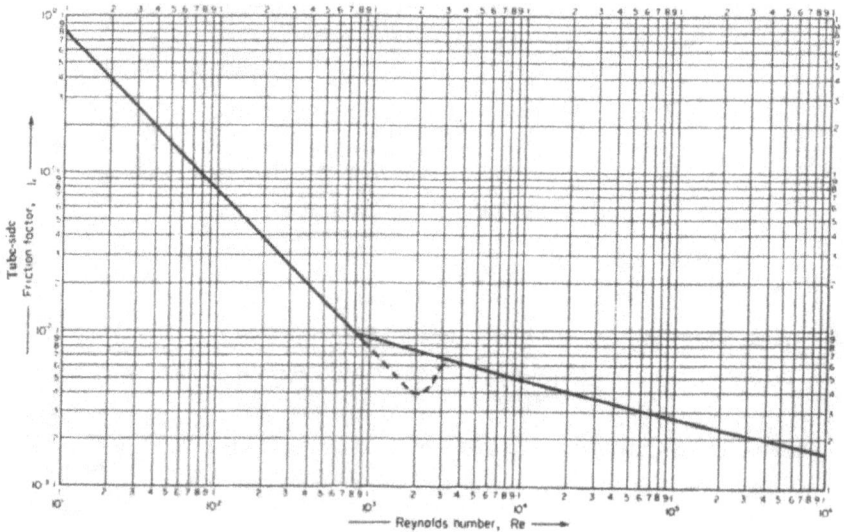

3.3.5.2: Pressure Drop outside Heat Exchanger Tubes

Flow outside the tubes, in heat exchangers, is often at right angles, to the tubes, which, in themselves, are, generally, arranged in rows, referred to as tube banks. Although no single expression will be general enough for estimating the shell side pressure drop in heat exchangers because of the many configurations in which these tubes can be arranged, the most general form, with expansion and contraction losses negligible ($K_{expansion} = 0 = K_{contraction}$), is given by Kays and London (1958) as

$$\Delta P = u_1 \frac{G^2}{2} \left[\left(1 + \sigma^2 \right) \left(\frac{u_2}{u_1} - 1 \right) + f \left(\frac{S}{S_t} \right) \cdot \left(\frac{u_m}{u_t} \right) \right] \qquad (3.47)$$

$$\underbrace{\qquad\qquad}_{flow\ acceleraton} \qquad \underbrace{\qquad\qquad}_{loss\ in\ tubes}$$

An explicit version of the above equation is, if entrance and exit losses are neglected because they are, usually, negligible,

$$\Delta P = 4 \left(\frac{D}{d_e} \right) \cdot \left(\frac{R}{\rho u^2} \right)_0 \left(\rho_0\, u_0^2 \right) \left(N_B + 1 \right) \qquad (3.48)$$

where N_B is the number of baffles, D is the inside diameter of the shell and d_e is the hydraulic mean, or equivalent, diameter given as

$$d_e = \frac{4}{\pi d_0} \cdot \left(p_t^2 - \frac{\pi d_0^2}{4} \right) \quad for\ square\ pitch\ arrangement$$

$$= \frac{8}{\pi d_0} \left(0.435 p_t^2 - \frac{\pi d_0^2}{8} \right) \quad for\ triangular\ pitch\ arrangement \quad (3.49)$$

and p_t and d_0 are the tube pitch and outside diameter, respectively. Also

$$\left(\frac{R}{\rho u^2} \right)_0 = 0.17 Re_0^{-0.18} \qquad (3.50)$$

$$Re_0 = \frac{G_s\, d_e}{\mu} \qquad (3.51)$$

$$G_s = \frac{W_s}{S_s} \qquad (3.52)$$

$$S_s = \frac{(p_t - d_0) D L_B}{p_t} \qquad (3.53)$$

W_s is the mass flow rate per unit time on the shell side and L_B is the baffle spacing. There is always leakage of shell side fluid through the spaces between tubes and baffle plates (see Figure 3.27). Though this leakage could contribute to increased heat transfer, it, also, contributes to increased pressure drop in shell side fluid flow. Sinnott (1983) identified five types of leakage and by-pass streams in shell fluid flow in a baffled shell and tube heat exchanger and described the various methods for predicting the consequent heat transfer and pressure drop in such situations.

There are three main methods, however, in general use for determining the pressure drop in shell side fluid flow. These are the so called "Bulk Flow Analysis Method" typified by Kern's method, the "Detailed Stream Analysis Method", typified by Tinker's method and its subsequent and propriety modifications, and the "Semi-Analytical Method" of Bell. These methods have been described in some detail by Sinnott (1983.

Briefly, however, Kern's method gives the pressure drop in terms of j_f factors and obtains

$$\Delta P = 8 j_f \left(\frac{D}{d_e}\right)\left(\frac{L}{L_B}\right)\left(\frac{\rho u_s^2}{2}\right)\left(\frac{\mu}{\mu_w}\right)^{-0.14} \tag{3.54}$$

where L is the length of tube and j_f is estimated from graphs of j_f versus Re (see Fig. 3.26). Note that $\dfrac{L}{L_B} = (N_B + 1)$ thus making equations (3.48) and (3.54) almost equivalent.

In Bell's method, the total pressure drop is the sum of pressure drops at the two end zones of the exchanger, at the (N_B - 1) crossflow zones and at the N_B window zones (see Fig. 3.28). That is

$$\Delta P = 2\,\Delta P_{end\,zone} + \Delta P_{cross\,flow}\,(N_B - 1) + N_B\,\Delta P_w \tag{3.55}$$

where the cross - flow pressure drop is given by

$$\Delta P_{cross\,flow} = \Delta P_{itb}.F_B.F_L \tag{3.56}$$

and the pressure drop for an ideal tube bank, ΔP_{itb}, is given by

$$\Delta P_{itb} = 8 j_f\, N_{CV}\left(\frac{\rho u_s^2}{2}\right)\left(\frac{\mu}{\mu_w}\right)^{-0.14} \tag{3.57}$$

while the by-pass correction factor, F_B, is given by

$$F_B = \exp\left[-\alpha\left(\frac{S_b}{S_s}\right)\left(1 - 2\frac{N_s}{N_{CV}}\right)^{\frac{1}{3}}\right] \tag{3.58}$$

N_s = number of sealing strips encountered by the by-pass stream in the crossflow zone $\leq 0.5 N_{CV}$

N_{CV} = number of constrictions and tube rows encountered in the crossflow section

S_b = clearance area between the tube bundle and shell

S_s = maximum area for crossflow, as given by equation (3.53)

α = 5.0 for laminar flow (Re < 100)

$$= \quad 4.0 \text{ for transition and turbulent flow (Re} > 100)$$

$$F_L = 1 - \beta_L \left(\frac{S_{ttb} + 2 S_{sb}}{S_L} \right) \tag{3.59}$$

where

S_{ttb} = the tube to baffle clearance area per baffle
S_{sb} = shell to baffle clearance area per baffle
S_L = total leakage area = $S_{ttb} + S_{sb}$
β_L = a factor obtained from Fig 3.29

The end zone pressure drop is

$$\Delta P_{end \, zone} = \Delta P_i \left[\frac{N_{wv} + N_{CV}}{N_{CV}} \right] F_B \tag{3.60}$$

where N_{wv} = the number of restrictions for crossflow in the window zone, approximately equal to the number of tube rows. Note that

$$N_{CV} = \frac{D_b - 2 H_b}{P_t} \tag{3.61}$$

$$N_{wv} = \frac{H_b}{P_t} \tag{3.62}$$

$$H_b = \frac{D_b}{2} - D(0.5 - B_C) \tag{3.63}$$

where D_b is the tube bundle diameter and H_b is the height from the baffle chord to the top of the tube bundle. B_C is the baffle cut, as a fraction, $p_t = p_t$ for a square pitch and $p_t = 0.87 p_t$ for an equilateral triangular pitch. The window zone pressure drop, ΔP_w, is given by

$$\Delta P_w = F_L \left(2 + 0.6 N_{wv} \right) \left(\frac{\rho u_z^2}{2} \right) \tag{3.64}$$

where

u_z = the geometric mean velocity = $\sqrt{u_w . u_s}$

$u_w = \dfrac{W_s}{\rho S_w}$, the velocity in the window zone based on the window area less the area occupied by the tubes, S_w.

$$S_w = \frac{\pi \left(D^2 R_A - N_w d_0^2 \right)}{4} \tag{3.65}$$

$$N_w = N_t \; x \; R_A \tag{3.66}$$

N_t is the total number of tubes, N_w the number in a window zone and R_A is the ratio of the bundle cross-sectional area in the window

113

zone to the total bundle cross-sectional area. Bell's method will predict more accurate values of pressure drop than Kern's method but is said not to be sufficiently accurate for designs in which the allowable pressure drop is the critical design parameter.

Peters & Timmerhaus (1985) give the expression

$$\Delta P_0 = \frac{N_R \, B_0 \, 2 \, f_0 \, G_s^2}{\rho_0} \tag{3.67}$$

where f_0 is a special friction factor for shell side flow and N_R, number of rows of tubes across which shell side fluid flows. B_0 is a correction factor to account for flow reversal, re-crossing of tubes and variation in cross-section. For unbaffled tubes, B_0 can be taken to be equal to 1 while, in other cases, B_0 can be equated to the number of tube crosses.

$$f_0 = b_0 \left(\frac{D G_s}{\mu_f} \right)^{-0.15} \tag{3.68}$$

For staggered tubes

$$b_0 = 0.23 + \frac{0.11}{\left(x_T - 1 \right)^{1.08}} \tag{3.69}$$

For tubes in line

$$b_0 = 0.044 + \frac{0.08 x_L}{\left(x_T - 1 \right)^{0.43 + 1.13 / x_L}} \tag{3.70}$$

where x_T is the ratio of the pitch, transverse to flow, to the tube diameter and x_L is the ratio of the pitch, parallel to flow, to the tube diameter.

Fig. 3.26: Friction Factor, j_f, as a function of Reynolds Number for Flow outside Tubes (Sinnott, 1983)

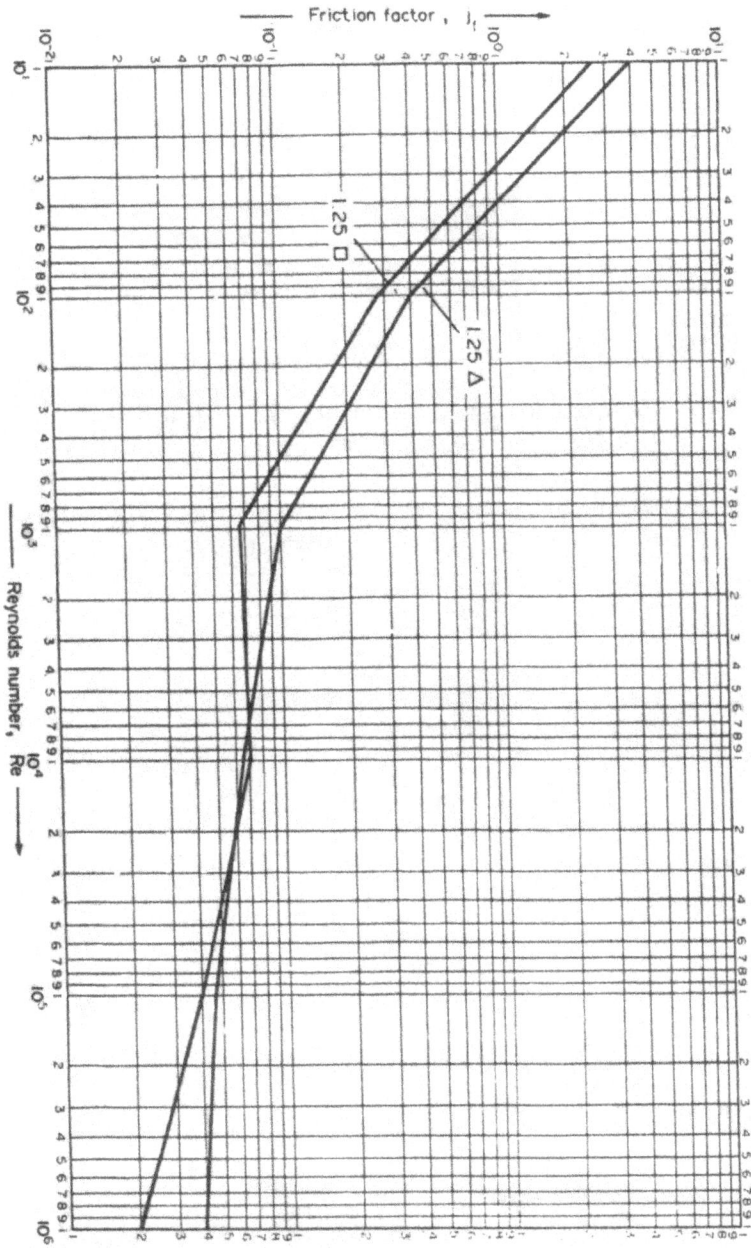

115

Fig. 3.27: By-pass and Leakage Streams in a Shell and Tube Heat Exchanger (Sinnot, 1983)

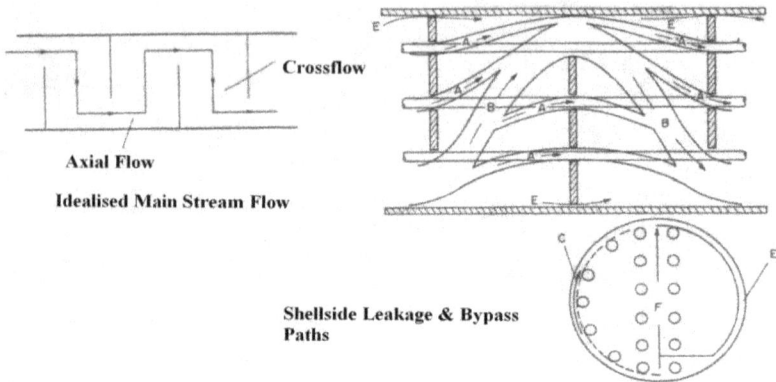

Crossflow

Axial Flow

Idealised Main Stream Flow

Shellside Leakage & Bypass Paths

Fig. 3.28: Clearances and Flow Areas in the Shell Side of a Shell and Tube Heat Exchanger (Sinnot, 1983)

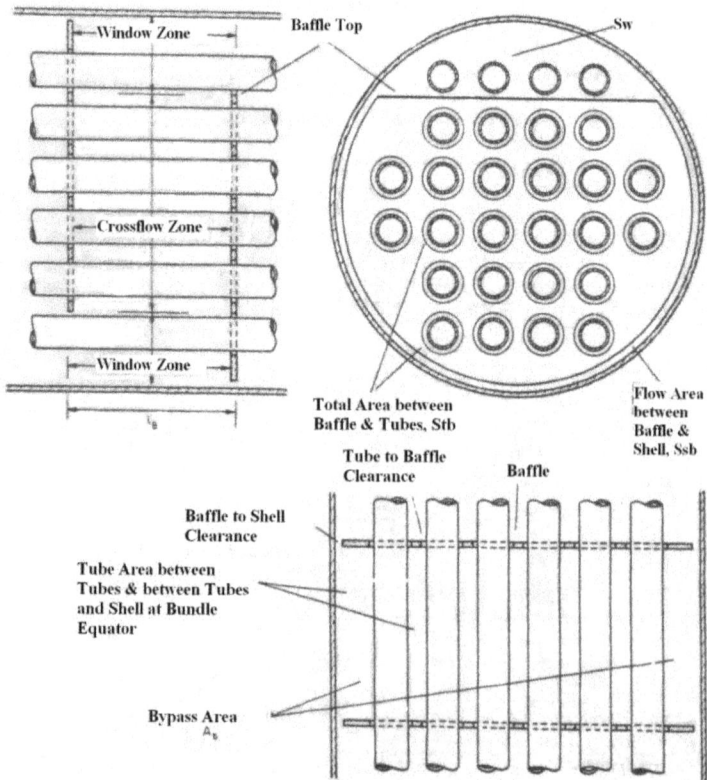

Window Zone

Baffle Top

Sw

Crossflow Zone

Window Zone

Total Area between Baffle & Tubes, Stb

Flow Area between Baffle & Shell, Ssb

Tube to Baffle Clearance

Baffle

Baffle to Shell Clearance

Tube Area between Tubes & between Tubes and Shell at Bundle Equator

Bypass Area
A_b

116

3.4: The Design and Analysis of Heat Exchangers

The two major objectives in heat exchanger design are

a. to calculate either the surface area, S, the heat transferred, Q or terminal temperatures, T_{ij} or ΔT.
b. to design for the lowest possible capital and operating cost.

3.4.1: Estimating S, Q or ΔT in Direct Type Heat Exchangers

Two common methods, in the public domain, used for estimating S, Q or ΔT in direct type heat exchangers are the log mean temperature difference method and the effectiveness-NTU method. Other methods which are, usually, refinements or advances on these are, however, often proprietary to particular manufacturers and designers of heat exchangers. Commercial design methods are , generally, computerized and marketed as heat exchanger design software.

3.4.1.1: <u>The Log Mean Temperature Difference (LMTD) Approach</u>

In this method, the exchanger surface area, for a single direct type heat exchanger, is estimated as follows. The heat transferred, or to be transferred, Q_T, is estimated from the stream for which there is complete information on entrance and exit temperatures, T_{ij}, fluid heat capacity, Cp, and fluid mass flowrate, F. There will always be one stream for which complete information is available or given. Thus, for this stream, from equation (3.5)

$$Q_T = F\,\overline{Cp}\,(\theta_{out} - \theta_{in}) = F\,(H_{out} - H_{in}) \qquad (3.42)$$

It is often possible to obtain, by a heat energy balance, the unkown exit or entrance temperatures in the other fluid stream. These enable us to determine the log mean temperature difference in the proposed heat exchanger, from equation (3.15)

$$\varphi_{lm} = \frac{\varphi_1 - \varphi_2}{\ln\left(\dfrac{\varphi_1}{\varphi_2}\right)} \qquad (3.43)$$

With all the terminal temperatures of the proposed heat exchanger known, the mean temperatures, required for evaluating the film heat transfer coefficients on both sides of the heat exchanger, can, now, be estimated. These enable computation of the film heat transfer coefficients from the appropriate dimensionless correlations. With these, the overall heat transfer coefficient, U, based on either side of fluid flow, can be evaluated as in equation (2.17). That is

$$\frac{1}{U_o\,A_o} = \frac{1}{U_i\,A_i} = \left[\frac{1}{A_i h_i} + \frac{x_w}{2 k_w\,A_{lm}} + \frac{1}{A_o\,h_o}\right] \qquad (3.44)$$

The subscripts, *i, o* and *w* represent, respectively, the inside, outside and wall of the tubes while *k* is the thermal conductivity of the tube material. The required surface area of the heat exchanger is given, from equation (3.16), by

$$S_o = \frac{Q_T}{U_o\,\varphi_{lm}} \quad or \quad S_i = \frac{Q_T}{U_i\,\varphi_{lm}} \qquad (3.45)$$

It should be noted that this design is for the so called clean surface heat exchanger. In practice, as the heat exchanger is used, the surfaces of the tubes of the exchanger become covered with deposits, called fouling, thereby increasing thermal resistances in the exchanger. These deposits may be silt or sediment from fluids carrying suspended solids, scale from crystallization or similar processes or film from polymerization in the fluids exchanging heat energy.

The thermal resistances offered by these deposits are, usually, predicted from experience and are expressed in terms of fouling factors. A fouling factor, R, is the inverse of the film heat transfer coefficient of the fouling substance. Thus, if R_o and R_i are the fouling factors on the outside and inside of the tubes, respectively, the overall heat transfer coefficient with fouling, U_F is related to the clean overall heat transfer coefficient, U_C, as

$$\frac{1}{U_F} = \frac{1}{U_C} + R_i + R_o \qquad (3.46)$$

Typical values of R are shown in Table 3.6

The log mean temperature difference was derived on the assumption that U is contant throughout the length of the exchanger. To account for the fact that this is not always true, especially, for multipass heat exchangers, it is often multiplied by a temperature correction factor, F, usually less than or equal to 1. With these corrections equation (3.45) now becomes

$$S_{oc} = \frac{Q_T}{U_{oc}\,\varphi_{lm}\,F} \quad or \quad S_{ic} = \frac{Q_T}{U_{ic}\,\varphi_{lm}\,F} \qquad (3.47)$$

The extra subscript, *c*, denotes the corrected value. When the temperature correction factor, F, is inadequate due to greater variation of U than can be corrected for by this factor and where it is possible, the exchanger can be divided into j sections of constant U_j and Q_j. The area for each section, S_j, is computed from

$$S_j = \frac{Q_j}{\left(U_{2j}.\Delta T_{1j} - U_{1j}.\Delta T_{2j}\right)} \ln\left(\frac{U_{2j}.\Delta T_{1j}}{U_{1j}.\Delta T_{2j}}\right) \tag{3.48}$$

Table 3.6: Experience Values of Heat Exchanger Fouling Factors (Sinnot, 1983)

Fouling Substance	Fouling Factor, m^2K/W
River water	0.00008 – 0.0003
Sea water	0.0003 – 0.001
Cooling tower water	0.00017 – 0.0003

Fouling Substance	Fouling Factor, m^2K/W
Town water (soft)	0.0002 – 0.0003
Town water (hard)	0.0005 – 0.001
Styeam condensate	0.0002 – 0.00067
Steam (oil free)	0.0001 – 0.00025
Steam (oil, traces)	0.0.0002 – 0.0005
Air and industrial gases	0.0001 – 0.0002
Flue gases	0.0002 – 0.0005

Situations arise in which it is possible to express h, the film heat transfer coefficient, as a function of some temperature difference. In such cases, the procedure developed by Sargent (1966) can be used. In this procedure and for any heat exchanger we may represent it as shown below. T_t is the tube temperature. Heat transfer is from the fluid flowing outside the tubes to that flowing inside.

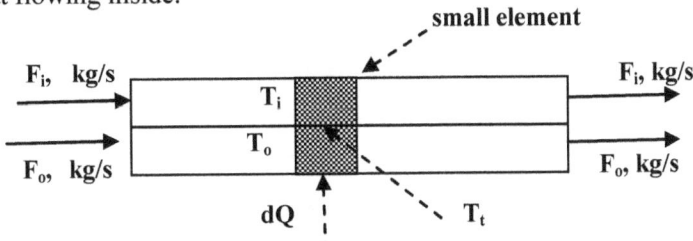

A heat energy balance on the element gives

$$F_i \, Cp_i \, dT_i = (T_t - T_i)\, h_i \, dS_i = (T_o - T_t)\, h_o \, dS_o \tag{3.49}$$

If we define

$$\psi = T_o - T_i \quad and \quad \phi = T_o - T_t \tag{3.50}$$

equation (3.49) becomes

$$-F_i \, Cp_i \, d\psi = (\psi - \phi)\, h_i \, dS_i = \phi\, h_o \, dS_o \tag{3.51}$$

119

Since S_i and S_o are related by their equivalent diameters,

$$\frac{S_i}{S_o} = \frac{D_i}{D_o} \quad or\; d\,S_i = \frac{D_i}{D_o} d\,S_o \tag{3.52}$$

Equation (3.51) with equation (3.52) becomes

$$- F_i\, Cp_i\, d\psi = (\psi - \phi)\, h_i \frac{D_i}{D_o} d\,S_o = \phi h_o\, d\,S_o \tag{3.53}$$

This enables us to obtain a value for ψ as

$$\psi = \phi\left(1 + \frac{h_o\, D_o}{h_i\, D_i}\right) \tag{3.54}$$

Using equation (3.54) in equation (3.51) or (3.53)

$$- F_i\, Cp_i\, d\phi\left(1 + \frac{h_o\, D_o}{h_i\, D_i}\right) = \phi h_o\, d\,S_o$$

That is

$$S_{oT} = \int_0^{S_oT} d\,S_o = -\int_0^{S_oT} F_i\, Cp_i \left(1 + \frac{h_o D_o}{h_i\, D_i}\right)\frac{d\phi}{\phi h_o} \tag{3.55}$$

Since h_o is known as a function of ϕ equation (3.55) can be solved, usually, with a trial and error procedure.

For multi-pass and cross-flow shell and tube heat exchangers, where the flow is not strictly parallel nor truly co-current or counter-current, the above methods become difficult to use and the *temperature correction factor method* has to be resorted to. Several authors have proposed formulae for the computation of F, using this method. One of such formulae is given as

$$F = \frac{\frac{\sqrt{(R^2+1)}}{(R-1)}\ln\frac{1-X}{1-R.X}}{\ln\left[\frac{\frac{2}{X}-1-R+\sqrt{R^2+1}}{\frac{2}{X}-1-R-\sqrt{R^2+1}}\right]} \tag{3.56}$$

where

$$R = \frac{T_{h,in} - T_{h,out}}{T_{c,out} - T_{c,in}},$$

$$P = Temperatue\; Efficiency = \frac{T_{c,out} - T_{c,in}}{T_{h,in} - T_{c,in}} \tag{3.57}$$

$$X = \cfrac{1 - \left(\cfrac{R.P-1}{P-1}\right)^{\frac{1}{N_{shell}}}}{R - \left(\cfrac{R.P-1}{P-1}\right)^{\frac{1}{N_{shell}}}} \qquad (3.58)$$

Before the widespread use of digital computers, graphs were mainly used to estimate these values of F. These graphs were available only in proprietary company handbooks or design data books, sometimes simplified for ease of use. Typical of such charts are those for a 1-2 and 2-4 multipass heat exchangers, shown below. Graphical estimation is, however, not as accurate as numerical computation.

Figure 3.25: Temperature Correction Factors for a 1-2 Heat Exchanger One Shell Pass, Two or Even Multiples of Tube Passes (Sinnott, 1983)

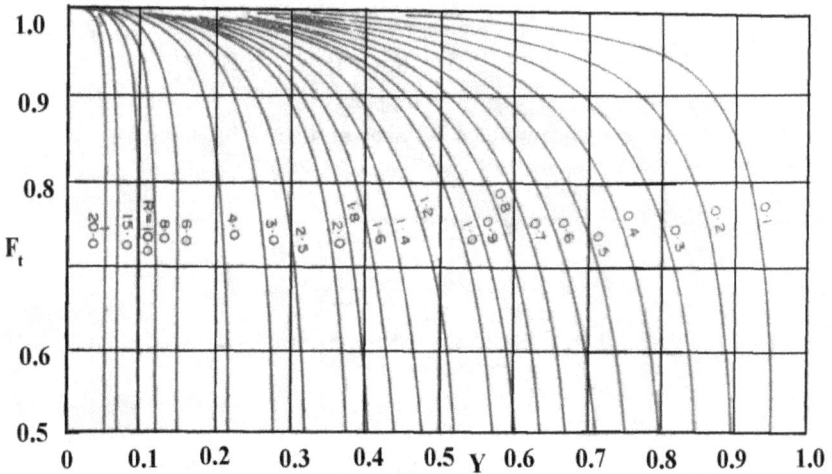

Schematics for a 1-2 Shell and Tube Exchanger

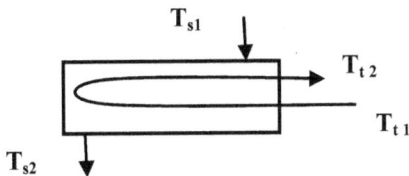

121

Figure 3.26: Temperature Correction Factors for a 2-4 Heat Exchanger Two Shell Passes, Four or Multiples of Tube Passe (Sinnott, 1963)

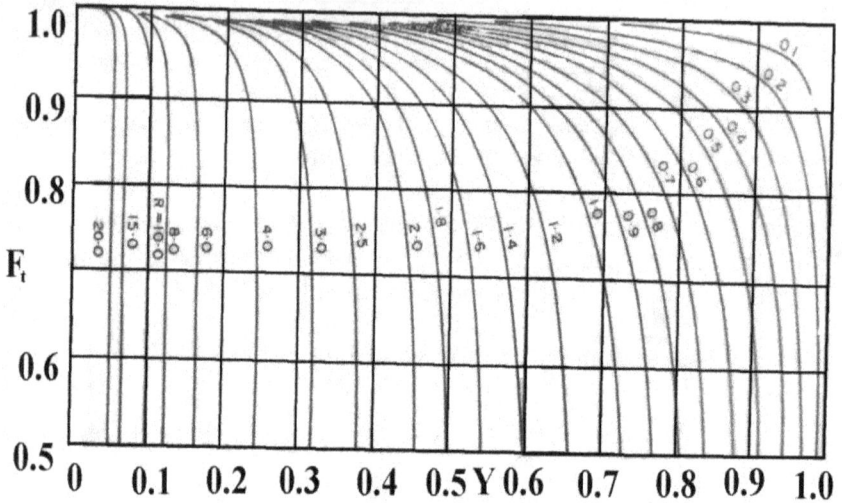

Schematics for a 2-4 Shell and Tube Exchanger

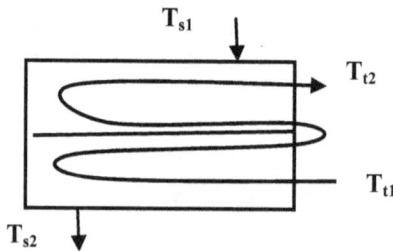

Computation of Y and R for both Charts

$$Y = \frac{T_{t,2} - T_{t,1}}{T_{s,1} - T_{t,1}} . \qquad R = \frac{T_{s,1} - T_{s,2}}{T_{t,2} - T_{t,1}}$$

3.4.1.2: Effectiveness - NTU Approach

This procedure was developed in response to the urgent need for compact, efficient and lightweight heat exchangers for the U.S. space program in the 1960s. Some of the early pioneering work was done at Stamford University, Connecticut, USA. Spin-offs of this technology have multiplied to other fields in electronics, computers, automobiles, and other domestic and office appliances.

In its early days, this approach depended on experimentally determined charts for all possible and conceivable compact surface configurations. These led to mathematical derivations or confirmations of the relationships between effectiveness and the number of transfer units. Tables 3.2 to 3.5 list some of these ε –NTU relations, without the charts, for a few configurations. Figure 3.30 illustrates one of the ε –NTU charts, this one, for a single pass counter current heat exchanger.

Most of the configurations were intended to be made, in cost efficient manufacturing processes, from lightweight corrugated aluminium or steel sheets overlaid on each other in co-current, countercurrent of crossflow arrangements. A few of these are illustrated in Figures 3.27, 3.28 and 3.29.

The problem of heat exchanger design or performance evaluation then becomes that of selecting, from the array of configurations, which one matched the specifications of one's heat exchanger design or performance problem. To do this, one needed to evaluate

i.   the ε or the NTU and $\mu = C_{min}/C_{max}$ from the information one had about one's system

ii. the equivalent NTU or ε, corresponding to that ε or NTU on the chart for the configuration chosen, for the same μ

iii. the exchanger surface as $S = \dfrac{NTU \cdot C_{min}}{U}$

iv. the heat transferred as $Q = C_{min} \left(\theta_{h,in} - \theta_{c,in}\right). \ \varepsilon$

v.   terminal temperatures from $Q = C_h \left(\theta_{h,in} - \theta_{h,out}\right)$ or   from
$Q = C_c \left(\theta_{c,out} - \theta_{c,in}\right)$

Figure 3.27: One Pass of a Compact Heat Exchanger Configuration (Triangular Channels)

Figure 3.28: One Pass of a Compact Heat Exchanger Configuration (Rectangular Channels)

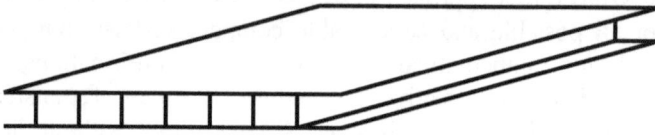

These one-pass structures can be stacked on top of each other in co-current, countercurrent or crossflow.

Figure 3.29: End and Side View Schematics of a Multi Tube Pass Plate/Sheet Finned Compact Heat Exchanger

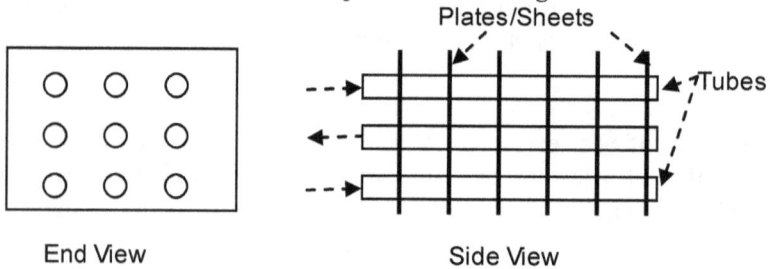

End View   Side View

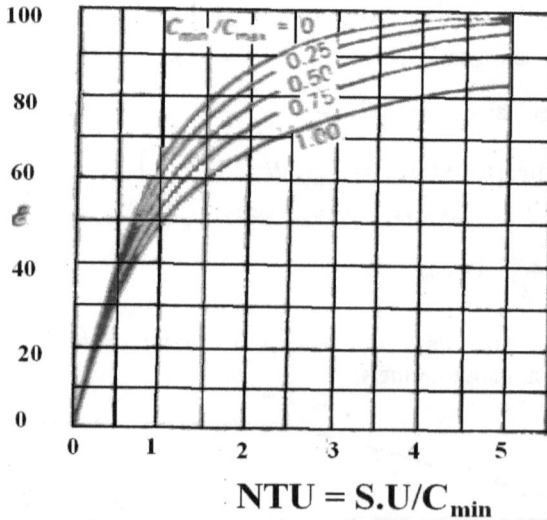

$$NTU = S.U/C_{min}$$

Figure 3.30: ε–NTU Chart for a Countercurrent Heat Exchanger (Kays & London, 1964)

WORKED EXAMPLES

Example 3.1

A certain oil, with specific heat capacity of 1.884 kJ/kg.K, is to be preheated from 301.62 K to 340 K in a heat exchanger, using a steam condensate at 350 K which is further cooled to 315 K. If the oil flowrate is 7.7 kg/s while the stream condensate, with specific heat capacity of 4.187 kJ/kg.K, flows at 3.8 kg/s, determine the effectiveness of this heat exchanger.

Answer

The effectiveness of a heat exchanger is given as

$$\varepsilon = \frac{Q_{actual}}{Q_{max}} = \frac{C_c}{C_{min}} \cdot \frac{(T_{c,out} - T_{c,in})}{(T_{h,in} - T_{c,in})} \quad if \quad C_c > C_h$$

$$= \frac{C_h}{C_{min}} \cdot \frac{(T_{h,in} - T_{h,out})}{(T_{h,in} - T_{c,in})} \quad if \quad C_c < C_h \qquad (1)$$

The heat capacity, C, is given by
$$C = flow\,rate \times specific\,heat\,capacity \qquad (2)$$

In this problem

$$C_h = 3.8 \times 4187 \frac{kg}{s} \cdot \frac{J}{kg.K} = 159106, \frac{W}{K}$$

$$C_c = 7.7 \times 1884 \frac{kg}{s} \cdot \frac{J}{kg.K} = 145068, \frac{W}{K}$$

$$C_c = C_{min} < C_h \qquad (3)$$

$$Q_{actual} = 7.7 \times 1884 \times (340 - 301.62) = 3.8 \times 4187 \times (350 - 315), \frac{kg}{s} \cdot \frac{J}{kg.K} \cdot \frac{K}{1}$$

$$= 556771 \approx 55687 1 JW$$

$$Q_{max} = C_{min} (T_{h,in} - T_{c,in}) = 145068 \times (350 - 301.62) = 70183898, \frac{W}{K} \cdot \frac{K}{1}$$

$$\varepsilon = \frac{Q_{actual}}{Q_{max}} = \frac{556771}{70183898} = 0.793 \quad Ans.$$

$$Note\,also \quad \varepsilon = \frac{C_c}{C_{min}} \cdot \frac{(T_{c,out} - T_{c,in})}{(T_{h,in} - T_{c,in})} = \frac{145068}{145068} \cdot \frac{(340 - 301.62)}{(350 - 301.62)} = 0.793 \quad Ans.$$

Example 3.2

Kerosene, with a specific heat capacity of 2.094 kJ/kg.K, is to be preheated with a steam condensate which enters the preheater at 93 C with a flowrate of 0.14 kg/s. The kerosene enters at 27 C at the rate of 0.32 kg/s. If the exit temperature of the kerosene is not to exceed 54C. determine the required surface area of the preheater using the log mean temperature difference approach. Take the specific heat capacity of condensed steam as 4.187 kJ/kg.K and the overall heat transfer coefficient, U, based on the condensate side, as 1477 W/m².K

Answer

The fluid flow directions and terminal temperatures are as illustrated below.

a) Counter-current flow

Kerosine in, 27 C ──────────────────▶ Kerosine out, 54 C

Condensate out,◀────────────────── Condensate in,
$T_{h,out}$ C   93 C

b) Co-current flow

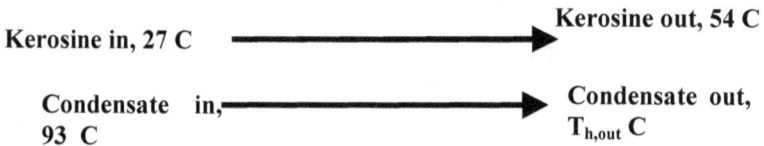

Kerosine in, 27 C ──────────────────▶ Kerosine out, 54 C

Condensate in,──────────────────▶ Condensate out,
93 C $T_{h,out}$ C

Since the exit temperature of condensate was not given, it has to be calculated from a heat energy balance in which the heat energy required to heat up the kerosene is equal to that lost by steam condensate in doing so. The total heat energy transferred, Q_T, is

$$Q_T = 0.32 x 2094 x (273 + 54 - 273 - 27), \frac{kg}{s}.\frac{J}{kg.K}.\frac{K}{} = 1809216, W \quad (1)$$

By a heat energy balance

$$1809216 = 0.14 x 4187 x (273 + 93 - 273 - T_{h,out})$$

$$T_{h,out} = 366 - \frac{1809216}{58618} = 335.14 \quad (2)$$

Since 27 C = 300 K, 54 C = 327 K and

$$\varphi_{lm} = \frac{\varphi_1 - \varphi_2}{\ln\left(\dfrac{\varphi_1}{\varphi_2}\right)} \qquad\qquad from\,(3.43)$$

For counter-current flow,

$$\varphi_{lm} = \frac{\varphi_1 - \varphi_2}{\ln\left(\dfrac{\varphi_1}{\varphi_2}\right)} = \frac{(335.14-300)-(366-327)}{\ln\dfrac{(335.14-300)}{(366-327)}} = \frac{-3.86}{-0.104} = 37.12K \qquad (2)$$

For co-current flow,

$$\varphi_{lm} = \frac{\varphi_1 - \varphi_2}{\ln\left(\dfrac{\varphi_1}{\varphi_2}\right)} = \frac{(366-300)-(335.14-327)}{\ln\dfrac{(366-300)}{(335.14-327)}} = \frac{57.86}{2.093} = 27.65K \qquad (3)$$

The required surface area, S, is, therefore, for counter-current flow,

$$S_o = \frac{Q_T}{U_o\,\varphi_{lm}} = \frac{1809216}{1477x37.12}, \frac{W}{W}.\frac{m^2.K}{K} = 0.330m^2 \quad from\,(3.45)\,Ans.$$

and for co-current flow,

$$S_o = \frac{Q_T}{U_o\,\varphi_{lm}} = \frac{1809216}{1477x27.65}, \frac{W}{W}.\frac{m^2.K}{K} = 0.443m^2 \quad from\,(3.45)\,Ans.$$

It can be seen that co-current flow requires more surface area for the same heat transfer than counter-current flow. It has been assumed that the temperature correction factor, F, is equal to 1 since flow is parallel and a 1-1 exchanger has been assumed.

Example 3.3

It is desired to cool 18.9 kg/s of ethylene glycol from 121 C to 104 C using toluene as a coolant in a 1-2 shell and tube exchanger. The toluene is heated from 27 C to 63 C. Determine the correction factor to be applied to the log mean temperature difference. What would the correction factor be for (a) a 2-4 shell and tube exchanger and (b) a single pass cross-flow exchanger, both fluids unmixed.

Answer

We shall assume that ethylene glycol is the tube fluid and toluene, the shell fluid. The terminal temperatures are then as shown in the figures below

For a 1-2 Shell and Tube Exchanger

$$T_{s1} = 27\ C$$

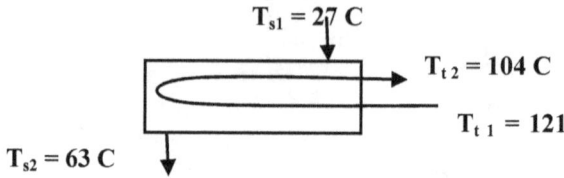

$$T_{t2} = 104\ C$$

$$T_{t1} = 121$$

$$T_{s2} = 63\ C$$

$$Y = \frac{T_{t,2} - T_{t,1}}{T_{s,1} - T_{t,1}} = \frac{104-121}{27-121} = 0.18. \quad R = \frac{T_{s,1} - T_{s,2}}{T_{t,2} - T_{t,1}} = \frac{27-63}{104-121} = 2.12$$

From Figure 3.25, F is estimated to be 0.98. Ans.

We can, also, estimate F from the formula of equation (3.56) given below.

$$R = \frac{T_{h,in} - T_{h,out}}{T_{c,out} - T_{c,in}} = \frac{121-104}{63-27} = 0.472. \quad P = \frac{T_{c,out} - T_{c,in}}{T_{h,in} - T_{c,in}} = \frac{63-27}{121-27} = 0.383$$

$$X = \frac{1 - \left(\dfrac{R.P-1}{P-1}\right)^{\frac{1}{N_{shell}}}}{R - \left(\dfrac{R.P-1}{P-1}\right)^{\frac{1}{N_{shell}}}} = \frac{1 - \left(\dfrac{0.472x0.383-1}{0.383-1}\right)^{1}}{0.472 - \left(\dfrac{0.472x0.383-1}{0.383-1}\right)^{1}} = \frac{-0.328}{-0.856} = 0.383$$

$$F = \frac{\dfrac{\sqrt{(R^2+1)}}{(R-1)}\ln\dfrac{1-X}{1-R.X}}{\ln\left[\dfrac{\dfrac{2}{X}-1-R+\sqrt{R^2+1}}{\dfrac{2}{X}-1-R-\sqrt{R^2+1}}\right]} = \frac{\dfrac{\sqrt{(0.472^2+1)}}{0.472-1}\ln\dfrac{1-0.383}{1-0.472x0.383}}{\ln\left[\dfrac{\dfrac{2}{0.383}-1-0.472+\sqrt{0.472^2+1}}{\dfrac{2}{0.383}-1-0.472-\sqrt{0.472^2+1}}\right]}$$

$$= \frac{-2.094x-0.284}{\ln\dfrac{4.8557}{2.6441}} = 0.978$$

The calculated result, though more accurate, is, certainly, more tedious than a graphical estimate. It has the advantage, however, of being capable of incorporation into a computer algorithm.

Suppose, however, that toluene was the tube fluid and ethylene glycol, the shell side fluid. The diagram of flow and terminal temperatures would, now, look like this.

128

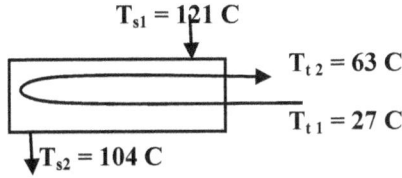

$$T_{s1} = 121\ C$$
$$T_{t2} = 63\ C$$
$$T_{t1} = 27\ C$$
$$T_{s2} = 104\ C$$

Then

$$Y = \frac{T_{t,2} - T_{t,1}}{T_{s,1} - T_{t,1}} \cdot = \frac{63 - 27}{121 - 27} = 0.38. \quad R = \frac{T_{s,1} - T_{s,2}}{T_{t,2} - T_{t,1}} = \frac{121 - 104}{63 - 27} = 0.47$$

From Figure 3.25, F is estimated to be, still, 0.98. Ans.

Also, F, from the formula of equation (3.56), is

$$R = \frac{T_{h,in} - T_{h,out}}{T_{c,out} - T_{c,in}} = \frac{121 - 104}{63 - 27} = 0.472. \quad P = \frac{T_{c,out} - T_{c,in}}{T_{h,in} - T_{c,in}} = \frac{63 - 27}{121 - 27} = 0.383$$

$$X = \frac{1 - \left(\frac{R.P - 1}{P - 1}\right)^{\frac{1}{N_{shell}}}}{R - \left(\frac{R.P - 1}{P - 1}\right)^{\frac{1}{N_{shell}}}} = \frac{1 - \left(\frac{0.472 x\, 0.383 - 1}{0.383 - 1}\right)^{1}}{0.472 - \left(\frac{0.472 x\, 0.383 - 1}{0.383 - 1}\right)^{1}} = \frac{-0.328}{-0.856} = 0.383$$

$$F = \frac{\frac{\sqrt{(R^2 + 1)}}{(R - 1)} \ln \frac{1 - X}{1 - R.X}}{\ln \left[\frac{\frac{2}{X} - 1 - R + \sqrt{R^2 + 1}}{\frac{2}{X} - 1 - R - \sqrt{R^2 + 1}}\right]} \cdot = \frac{\frac{\sqrt{0.472^2 + 1}}{0.472 - 1} \ln \frac{1 - 0.383}{1 - 0.472 x\, 0.383}}{\ln \left[\frac{\frac{2}{0.383} - 1 - 0.472 + \sqrt{0.472^2 + 1}}{\frac{2}{0.383} - 1 - 0.472 - \sqrt{0.472^2 + 1}}\right]}$$

$$= \frac{-2.094 x - 0.284}{\ln \frac{4.8557}{2.6441}} = 0.978$$

This example shows that fluid allocation has no influence on the log mean temperature correction factor.

For a 2-4 Shell and Tube Exchanger

Since fluid allocation does not affect the temperature correction factor, let us keep ethylene glycol still as the tube side fluid.
The terminal temperatures and fluid flow directions are illustrated below.

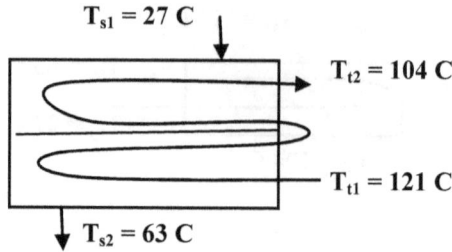

$$Y = \frac{T_{t,2} - T_{t,1}}{T_{s,1} - T_{t,1}} \cdot = \frac{104 - 121}{27 - 121} = 0.18. \quad R = \frac{T_{s,1} - T_{s,2}}{T_{t,2} - T_{t,1}} = \frac{27 - 63}{104 - 121} = 2.12$$

This gives a value for F, from Figure 3.26, of between 0.99 and 1.0. From the formula of equation (3.56), $N_{shell} = 2$.

$$R = \frac{T_{h,in} - T_{h,out}}{T_{c,out} - T_{c,in}} = \frac{121 - 104}{63 - 27} = 0.472. \quad P = \frac{T_{c,out} - T_{c,in}}{T_{h,in} - T_{c,in}} = \frac{63 - 27}{121 - 27} = 0.383$$

$$X = \frac{1 - \left(\dfrac{R.P - 1}{P - 1}\right)^{\frac{1}{N_{shell}}}}{R - \left(\dfrac{R.P - 1}{P - 1}\right)^{\frac{1}{N_{shell}}}} = \frac{1 - \left(\dfrac{0.472 x 0.383 - 1}{0.383 - 1}\right)^{\frac{1}{2}}}{0.472 - \left(\dfrac{0.472 x 0.383 - 1}{0.383 - 1}\right)^{\frac{1}{2}}} = \frac{-0.1523}{-0.6803} = 0.224$$

$$F = \frac{\dfrac{\sqrt{(R^2 + 1)}}{(R - 1)} \ln \dfrac{1 - X}{1 - R.X}}{\ln \left[\dfrac{\dfrac{2}{X} - 1 - R + \sqrt{R^2 + 1}}{\dfrac{2}{X} - 1 - R - \sqrt{R^2 + 1}}\right]} = \frac{\dfrac{\sqrt{0.472^2 + 1}}{0.472 - 1} \ln \dfrac{1 - 0.224}{1 - 0.472 x 0.224}}{\ln \left[\dfrac{\dfrac{2}{0.224} - 1 - 0.472 + \sqrt{0.472^2 + 1}}{\dfrac{2}{0.224} - 1 - 0.472 - \sqrt{0.472^2 + 1}}\right]}$$

$$= \frac{0.297}{0.299} = 0.993. \; Ans$$

Example 3.4

A heat exchanger, to be constructed from 21.2 mm ID, 25.4 mm OD, mild steel tubing, is to cool 6.95 kg/s of benzene from 339 K to 313 K, using 6.31 kg/s of water, available at 283 K. The overall heat transfer coefficient, based on the outer tube area, is 625 W/m^2.K. Determine the required heat transfer area for

      a).   co-current flow, single pass .
      b).   counter current flow, single pass.

Answer

For <u>co-current flow</u>, a single pass heat exchanger may be represented, schematically, as follows

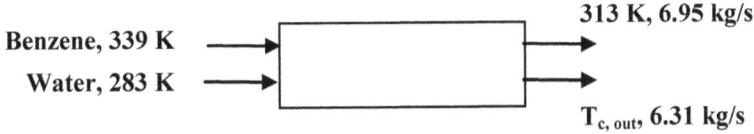

Benzene, 339 K →

Water, 283 K →

313 K, 6.95 kg/s

$T_{c, out}$, **6.31 kg/s**

Since the exit water temperature was not given, it has to be calculated from a heat energy balance. The specific heat capacities of benzene and water were not given. We can find out, from standard tables, that the specific heat capacity of water in this range of temperatures, is 4.187 kJ/kg.K. For benzene, at the mean temperature of *(339 + 313)/2 = 326 K.,* the specific heat capacity is estimated, by interpolation, from standard tables, as 1.8255 kJ/kg K.

The heat transferred, Q, based on benzene cooling, is, therefore,

$$Q = 6.95 x 1 8255 x (339 - 313), \frac{kg}{s}. \frac{J}{kg.K}. \frac{K}{} = 329,867.85, W \quad (1)$$

A heat energy balance, based on the heat transferred to the water stream, is

$$Q = 329,867.85, W = 6.31 x 4187 x (T_{c,out} - 283), \frac{kg}{s}. \frac{J}{kg.K}. \frac{K}{}$$

From which $T_{c,out}$ = 295.5 K. This enables us to evaluate the log mean temperature difference as

$$\Delta T_{lm} = \frac{(339 - 283) - (313 - 295.5)}{\ln \frac{(339 - 283)}{(313 - 295.5)}} = \frac{38.5}{1.1632} = 33.1 K \quad (2)$$

Since $Q = U.S_0.\Delta T_{lm}$. *F* where F = 1 for a single pass heat exchanger,

$$S_0 = \frac{Q}{U x \Delta T_{lm}} = \frac{329,867.85}{625 x 33.1 x 1}, \frac{W}{1}. \frac{m^2.K}{W}. \frac{1}{W} = 15.95 m^2 \quad Ans.$$

For <u>counter-current flow</u>, a single pass heat exchanger may, also, be represented, schematically, as

Benzene, 313 K

Water, 283 K

339 K, 6.95 kg/s

$T_{c, out}$, 6.31 kg/s

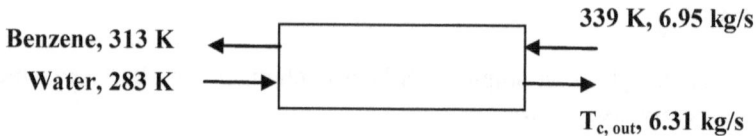

This time, the log mean temperature difference is

$$\Delta T_{lm} = \frac{(313-283)-(339-295.5)}{\ln\dfrac{(313-283)}{(339-295.5)}} = \frac{-13.5}{-0.3716} = 36.3\,K \qquad (3)$$

We can, thus, calculate S_0 as

$$S_0 = \frac{Q}{U \times \Delta T_{lm}} = \frac{329,867.85}{625 \times 36.3 \times 1}, \frac{W}{1} \cdot \frac{m^2.K}{W} \cdot \frac{1}{W} = 14.54 m^2 \quad Ans.$$

This problem shows that counter current flow requires less surface area, for the same duty, than co-current flow in a heat exchanger.

Example 3.5

Aniline is cooled from 325 K to 306 K in the inner pipe of a double pipe, single pass, heat exchanger. Cooling water flows counter-currently to the aniline, entering at 289 K and leaving at 301 K. The heat exchanger is constructed of a 18.9 mm ID, 22.2 mm OD copper tube, and jacketed with a 40.9 mm ID, 48.3 mm OD steel pipe. The velocity of the aniline is to be 1.5 m/s. Determine
  (a) the overall heat transfer coefficient, based on the outside area of the inner pipe
  (b) the required length of exchanger pipe

Answer

The exchanger may be represented as follows

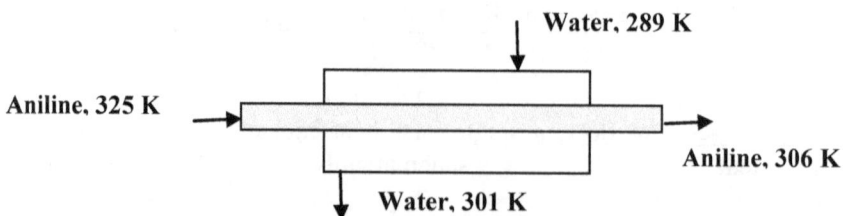

Water, 289 K

Aniline, 325 K

Aniline, 306 K

Water, 301 K

To find the overall heat transfer coefficient, we can use the Colburn equation for the internal flow of aniline

$$St = \frac{h}{\rho\, Cp\, u} = 0.023 Re^{-0.2}\, Pr^{-\frac{2}{3}}$$

$$for \quad Re > 10^4,\, 0.7 < Pr < 160,\, \frac{L}{D} > 60 \tag{1}$$

and, for external flow of water

$$j_h = \frac{0.351}{Re^{0.45}} \qquad for \quad 5{,}000 < Re < 10^6 \tag{2}$$

These formulae suggest that we have to evaluate the Re, Pr and L/D to be able to use equations (1) and (2).

For <u>aniline flow inside the tube</u>

Mean temperature = (325 + 306)/2 = 315.5 K.
At this temperature,

$\rho = 1005.5$ kg/m^3
$\mu = 2.49$ x 10^{-3}, kg/m.s (N.s/m^2)
$Cp = 2.061$ kJ/kg.K

$$M_t = \rho u = 10055 x 1.5, \frac{kg}{m^3}.\frac{m}{s} = 150825, \frac{kg}{m^2.s}$$

$k = 1.73$ x 10^{-4}, kW/m.K

The heat energy, Q, lost by aniline, is
$Q = \rho u\, A\, Cp\, \Delta T$

$$= 10055 x 1.5 x \pi x \frac{0.0189^2}{4} x 2061x (325-306), \frac{kg}{m^3}.\frac{m}{s}.\frac{m^2}{1}.\frac{J}{kg.K}.\frac{K}{1}$$

$$= 1657651 W \tag{3}$$

$$Re = \frac{\rho_i\, d_i\, u_i}{\mu_i} = \frac{10055 x 0.0189 x 1.5}{0.00249} = 114482 > 10^4 \tag{4}$$

$$Pr = \frac{Cp_i\, \mu_i}{k_i} = \frac{2061 x 0.00249}{0.173}, \frac{J}{kg.K}.\frac{kg}{m.s}.\frac{s.m.K}{J} = 29.6641 \tag{5}$$

From (1), (4) and (5)

$$St_i = \frac{h_i}{\rho_i\, Cp_i\, u_i} = 0.023 Re_i^{-0.2}\, Pr_i^{-\frac{2}{3}} = \frac{0.023}{(114482)^{0.2}\, x\, (29.664)^{\frac{2}{3}}}$$

$$= 3.70 x 10^{-4}$$

133

$$h_i = 3.70x10^{-4} \; x \, \rho_i \, Cp_i \, u_i = 3.70x10^{-4} \, x10055 \, x \, 206 \, 1x1.5, \frac{kg}{m^3}.\frac{J}{kg.K}.\frac{m}{s}$$

$$= 115014, \frac{W}{m^2.K} \qquad (6)$$

For <u>water flow outside the tube,</u>

Mean temperature = $(289 + 301)/2 = 295$ K.
At this temperature,

$\rho = 998.9$ kg/m^3
$\mu = 1.002$ x 10^{-3}, kg/m.s (N.s/m^2)
$Cp = 4.187$ kJ/kg.K
$k = 6.01$ x 10^{-4}, kW/m.K

The thermal conductivity of copper pipe at the mean temperature of $(315.5 + 295)/2 = 305.25$ K is, from standard tables, 401 W/m.K.

The equivalent diameter, d_e, has to be used and is given by

$$d_e = \frac{4 \, x \, wetted \; cross-sectional area}{wetted \; perimeter} = \frac{4x\left(\frac{\pi D_i^2}{4} - \frac{\pi d_0^2}{4}\right)}{\pi D_i + \pi d_0}$$

$$= D_i - d_0 = 0.0409 - 0.0222 = 0.0187m \qquad (7)$$

where D_i is the inner diameter of the steel pipe and d_0 is the outer diameter of the copper pipe.

The rate of cooling water flow is determined from a heat energy balance.

$$Q = m_w Cp \, \Delta T = m_w \, x \, 4187x(301-289), \frac{kg}{s}.\frac{J}{kg.K}.\frac{K}{1} \quad giving$$

$$m_w = \frac{1657651}{4187x(301-289)}, \frac{J}{s}.\frac{kg.K}{J}.\frac{1}{K} = 0.33\frac{kg}{s} \qquad (8)$$

The cross-sectional area, S_A, of the annulus, between the copper and steel tubes, is given by

$$S_A = \frac{\pi D_i^2}{4} - \frac{\pi d_0^2}{4} = \frac{\pi}{4}\left(0.0409^2 - 0.0222^2\right) = 9.271x10^{-4} \, m^2 \quad (9)$$

Note that this is different from the cross-sectional area based on the equivalent diameter ($S_{de} = 2.75$ x 10^{-4}, m^2).

$$M_A = \rho_w u_A, \frac{kg}{m^3} \cdot \frac{m}{s} \cdot \frac{m_w}{S_A} = \frac{0.33}{9.271x10^{-4}}, \frac{kg}{s} \cdot \frac{1}{m^2} = 355.95, \frac{kg}{s.m^2} \quad (10)$$

$$Re_0 = \frac{\rho_w d_e u_A}{\mu_w} = \frac{d_e M_A}{\mu_w}, \frac{m}{1} \cdot \frac{kg}{s.m^2} \cdot \frac{m.s}{kg}$$

$$= \frac{0.0187x355.95}{0.001002} = 664298 < 10^6 \; and \; greater than 5,000 \quad (11)$$

$$Pr_0 = \frac{Cp_w \, \mu_w}{k_w} = \frac{4187x0.001002}{0.601}, \frac{J}{kg.K} \cdot \frac{kg}{m.s} \cdot \frac{s.m.K}{J}$$

$$= 6.98 > 0.7 \; and \; less \; than 160 \quad (12)$$

From (2), (11) and (12)

$$St_0 = \frac{h_0}{\rho_w Cp_w u_A} = \frac{h_0}{M_A Cp_w} = \frac{0.351}{Re^{0.45} . Pr^{\frac{2}{3}}} = \frac{0.351}{(664298)^{0.45} x (6.98)^{\frac{2}{3}}}$$

$$= 1.83x10^{-3}$$

$$h_0 = 1.83x10^{-3} \; x M_A Cp_w = 1.83x10^{-3} x 355.95x4187, \frac{kg}{s.m^2} \cdot \frac{J}{kg.K} \quad The$$

$$= 2727.36, \frac{W}{m^2.K} \quad (13)$$

The overall heat transfer coefficient, based on the outside area of the inner pipe, U_0, is given by (from equation 2.20 or 2.21)

$$\frac{1}{U_0} = \frac{d_0}{h_i d_i} + \frac{d_0}{2k_t} \ln\left(\frac{d_0}{d_i}\right) + \frac{1}{h_0}$$

$$= \frac{0.0222}{11501 4 x 0.0189} + \frac{0.0222}{2x401} . \ln\left(\frac{0.0222}{0.0189}\right) + \frac{1}{2727.36}$$

$$= 0.0010212 7 + 0.0000044 6 + 0.0003666 6 + 0.00139239$$

That is, $U_0 = 718.19$ W/m^2.K.   Ans. $\quad (14)$

Since, for a clean surface, and single pass heat exchanger, F = 1

$$S_0 = \pi d_0 L = \frac{Q}{U_0 . \Delta T_{lm}} \; ,$$

$$\Delta T_{lm} = \frac{(325 - 301) - (306 - 289)}{\ln \dfrac{(325 - 301)}{(306 - 289)}} = \frac{7}{0.3448} = 20.30K \; and$$

$$L = \frac{S_0}{\pi d_0} = \frac{1}{\pi d_0} x \frac{Q}{U_0 . \Delta T_{lm}} = \frac{1}{\pi x 0.0222} x \frac{1657651}{71819x 20.30}, \frac{1}{m} \cdot \frac{W}{1} \cdot \frac{m^2.K}{W} \cdot \frac{1}{K}$$

135

That is $L = 16.30m.$ *Ans*

Example 3.6

12.6 kg/s of water (specific heat capacity = 4183 J/kg K) are to pass through a heat exchanger to raise its temperature from 60 C to 94 C, using a flue gas of specific heat capacity, 1005 J/kg K. The flue gas is available at 12.6 kg/s and 427 C. Determine the maximum heat transferable if the two streams flow co-currently. How does this compare to the actual heat transferred?

Answer

The system may be represented, schematically, as follows

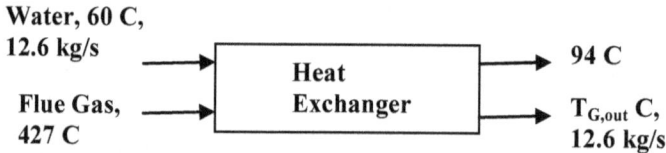

Water, 60 C,
12.6 kg/s ──────▶ ┌─────────────────┐ ──────▶ **94 C**
                     │   **Heat**   │
Flue Gas, ──────▶ │ **Exchanger**   │ ──────▶ **T$_{G,out}$ C,**
427 C   └─────────────────┘   **12.6 kg/s**

By a heat energy balance, the actual heat energy transferred to the water is

$$Q_w = m_w . Cp_w . \Delta T_w = 12.6 \times 4183 \times \{(273+94) - (273+60)\}, \frac{kg}{s}.\frac{J}{kg.K}.\frac{K}{1}$$

$$= 1,791,997.2 W \qquad\qquad (1)$$

The maximum heat energy transferable is obtained at that temperature in which Q_w is equal to Q_{FG}. That is

$$Q_\infty = m_{FG} . Cp_{FG} . (427 - T_\infty) = m_w . Cp_w . (T_\infty - 60) \quad or$$

$$12.6 \times 1005 \times \{427 - T_\infty\} = 12.6 \times 4183 \times \{T_\infty - 60\}, \frac{kg}{s}.\frac{J}{kg.K}.\frac{K}{1}$$

where Q_w, m_w, Cp_w, ΔT_w are, respectively, the actual heat energy transferred to the water, the mass flowrate, specific heat capacity and the temperature rise in the heated water stream while $Q_{FG}, m_{FG}, Cp_{FG}, T_\infty$ are the equivalent terms in the flue gas stream except for T_∞, Q_∞ which are the temperature and maximum heat transferrable between the incoming flue gas and cold water streams.

Thus we get that

136

$$T_\infty = \frac{\left(427 + \dfrac{4183}{1005} x\, 60\right)}{\left(1 + \dfrac{4183}{1005}\right)} = 131.09C \quad and$$

$$Q_\infty = 12.6\, x\, 1005 x\, (427 - 131.09) = 3,747,108.33\ W \quad Ans.$$

It can be seen that the maximum heat energy transferable is a little more than two times (2.09) the actual heat transferred. Q_∞ may, also, be determined graphically. This is more illustrative but less accurate than direct calculation. The data for the plot is tabulated as follows.

Table Showing the Computed Values of Q_w and Q_{FG}

T, C	Q_w, kW	Q_{FG}, kW
60	0	4647.321
70	527.058	4520.691
80	1054.116	4394.061
90	1581.174	4267.431
100	2108.232	4140.801
120	3162.348	3887.541
140	4216.464	3634.281
160	5270.58	3381.021
180	6324.696	3127.761
200		2874.501
250		2241.351
300		1608.201
350		975.051
400		341.901
450		-291.249

Q versus T Diagram for Co-Current Flow showing Q∞

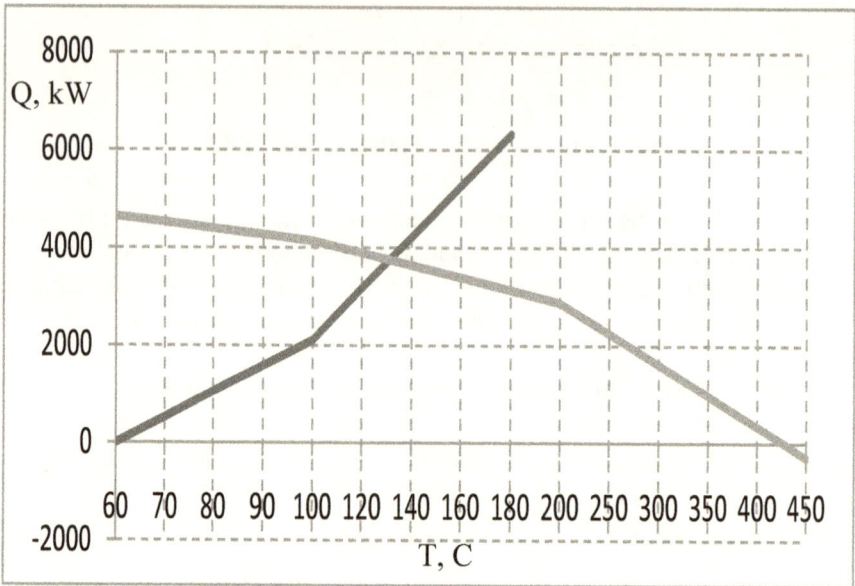

Example 3.7

Ethanol is to be condensed in the shell of an exchanger using cooling water whose temperature rises from 18 C to 27 C. The water flowrate is 1500 gal/h. If 19 mm OD x 16 SWG copper tubes are to be used, calculate the total length of tube required using the mean overall heat transfer coefficient. Compare your result with those obtained by the methods of sectioning the heat exchanger in a discrete or continuous manner. The heat transfer coefficient on the water side is 4.543 kW/m^2K whilst that for condensing ethanol is given by $h_0 = 3.89\theta^{1/4}$ where θ is the difference in temperature, in degrees K, between ethanol and the tube surface.

Answer

The system may be represented, schematically, as follows

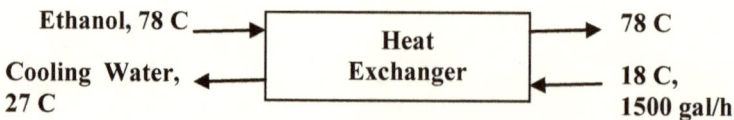

Ethanol, 78 C ⟶ | Heat Exchanger | ⟶ 78 C

Cooling Water, 27 C ⟵ | | ⟵ 18 C, 1500 gal/h

From standard Tables we obtain that, for water, Cp = 4.183 kJ/kg.K, density is 998 kg/m^3; for ethanol, the latent heat of condensation, λ, is 855 kJ/kg while for the copper tube, the thermal conductivity, k, is given as $k = 411.64 - 0.0875\ T$, W/m.K and T is in Kelvin. Also, for a 19mm OD x 16 SWG copper tube, the internal diameter is 15.875 mm.

The heat transferred, Q, based on the cooling water stream, is

$$Q_w = m_w \cdot Cp_w \cdot \Delta T_w$$

$$= 1500x\,4\,x\frac{1}{10^3}\,x\frac{1}{3600}\,x998x\,4183x\{(273+27)-(273+18)\},$$

$$\frac{gal}{h}\,x\frac{l}{gal}\,x\frac{m^3}{l}\,x\frac{h}{s}\,x\frac{kg}{m^3}\,x\frac{J}{kg.K}\,x\,K$$

$$= 62,619.51W \tag{1}$$

The log mean temperature difference is

$$\Delta T_{lm} = \frac{(78-27)-(78-18)}{\ln\dfrac{(78-27)}{(78-18)}} = \frac{-9}{-0.1625} = 55.39K \tag{2}$$

The temperature difference between ethanol and the tube surface, θ, (assumed to be at the temperature of the water in contact with it) varies from (78-18 = 60 C) at the cooling water entrance to (78-27 = 51) at the cooling water exit. The mean θ is, therefore $(51+60)/2 = 55.5K$

The overall heat transfer coefficient.based on the water side, U$_i$, is

$$\frac{1}{U_i} = \left[\frac{1}{h_i}+\frac{D_i}{2k}\ln\left(\frac{D_0}{D_i}\right)+\frac{D_i}{h_0\,D_0}\right] \qquad \textit{from (2.20)}$$

$$= \frac{1}{4543}+\frac{0.015875}{2(411.64-0.0875T_w)}\ln\left(\frac{0.019}{0.015875}\right)+\frac{0.015875}{0.019x3890(78-T_w)^{-0.25}}$$

$$= 0.000220+\frac{0.001426}{411.64-0.0875T_w}+0.000215(78-T_w)^{0.25} \tag{3}$$

Equation (3) may be simplified to

$$\frac{1}{U_i} = 0.000220+\frac{0.001426}{411.64-0.0875(78-T_w-78)}+0.000215(78-T_w)^{0.25}$$

$$= 0.000220+\frac{0.001426}{418.465-0.0875\theta}+0.000215\theta^{0.25}$$

$$= \frac{0.092062-0.000019\theta+0.001426+0.089970\theta^{0.25}-0.000019\theta^{1.25}}{418.465-0.0875\theta}$$

$$= \frac{0.093488 + 0.089970\theta^{0.25} - 0.0000199 - 0.0000199^{1.25}}{418465 - 0.08750} \quad (4)$$

At the mean temperature difference of 55.5 K

$$\frac{1}{U_i} = \frac{0.093488 + 0.089970 \times 55.5^{0.25} - 0.000019 \times 55.5 - 0.000019 \times 55.5^{1.25}}{418465 - 0.0875 \times 55.5}$$

$$= \frac{0.335123}{413.60875} \quad or \quad U_i = 1,23420 \frac{W}{m^2.K} \quad (5)$$

$Q_w = U_i \, S_i \, \Delta T_{lm} \, F$ so

$$S_i = \frac{Q_w}{U_i \times \Delta T_{lm}} = \frac{62,619.51}{1,23420 \times 55.39 \times 1} \cdot \frac{J}{s} \cdot \frac{s.m^2.K}{J} \cdot \frac{1}{K} = 0.92m^2 \quad (6)$$

Since $S_i = \pi D_i L$ then

$$L = \frac{S_i}{\pi \times D_i} = \frac{0.92}{\pi \times 0.015875} \cdot \frac{m^2}{1} \cdot \frac{1}{m} = 18.45m \quad Ans. \quad (7)$$

A mean value of U_i may, however, be found by integration rather than by evaluation at a mean temperature difference. In this case, using equation (4)

$$U_i = \frac{418465 - 0.08750}{0.093488 + 0.089970\theta^{0.25} - 0.0000199 - 0.0000199^{1.25}} \quad (8)$$

The mean U_i is then

$$\overline{U_i} = \frac{\displaystyle\int_{\theta_1}^{\theta_2} U_i \, d\theta}{\displaystyle\int_{\theta_1}^{\theta_2} d\theta}$$

$$= \frac{1}{\theta_2 - \theta_1} \cdot \int_{\theta_1}^{\theta_2} \frac{(418465 - 0.08750)d\theta}{0.093488 + 0.089970\theta^{0.25} - 0.0000199 - 0.0000199^{1.25}}$$

The term under the integration sign, I, can be integrated, numerically, using Weddle's rule, said to be more accurate than Simpson's rule. According to this rule, if $(\theta_2 - \theta_1)$ is divided into six equal intervals such that $h = (\theta_2 - \theta_1)/6$ then

$$\overline{U_i}(\theta_2 - \theta_1) = \frac{3h}{10} \cdot \left[I_{i0} + 5I_{i1} + I_{i2} + 6I_{i3} + I_{i4} + 5I_{i5} + I_{i6} \right] \quad (9)$$

The Is, computed with this value of h, are shown in the table below.

Ethanol Temp, C	Tube Temp, C	θ, C	I_i, W/m²K
78	27	51	1253.184
78	25.5	52.5	1246.656
78	24	54	1240.332
78	22.5	55.5	1234.2
78	21	57	1228.25
78	19.5	58.5	1222.472
78	18	60	1216.856

From the table and equation (9)

$$\overline{U}_i(60-51) = \frac{3x1.5}{10}.[1253184 + 5x1246656 + 1240332 + 6x12342]$$

$$+ \frac{3x1.5}{10}[122825 + 5x1222472 + 1216856] = 111102579$$

That is

$$\overline{U}_i = \frac{111102579}{9} = 12344731 \frac{W}{m^2.K} \tag{10}$$

Note how close it is to that in equation (5) and yet without the tedious computation.

Example 3.8

A light oil is to be cooled from 100 C to 70 C in a direct heat exchanger using water, available at 30 C. The oil flowrate is 0.5 kg/s while the available water rate is 0.92 kg/s. If the specific heat of the oil is 3.85 kJ/kg.K, and that of water, 4.18 kJ/kg.K, compare the effectiveness of this exchanger in co-current and counter-current flow, given that the overall heat transfer coefficient is 500 W/m².K. The surface area of the exchanger is 2.5 m².

Answer

The effectiveness of a heat exchanger is given as

$$\varepsilon = \frac{Q_{actual}}{Q_{max}} = \frac{C_c}{C_{min}}.\frac{(T_{c,out} - T_{c,in})}{(T_{h,in} - T_{c,in})} \quad if \quad C_c > C_h$$

$$= \frac{C_h}{C_{min}}.\frac{(T_{h,in} - T_{h,out})}{(T_{h,in} - T_{c,in})} \quad if \quad C_c < C_h \tag{1}$$

where the heat capacity, C, is given by

141

$$C = flow\,rate \times specific\,heat\,capacity \qquad (2)$$

The effectiveness of a heat exchanger is, also, given for

Co-current Flow as $\qquad \varepsilon = \dfrac{1 - e^{-NTU\,(1+\mu)}}{(1+\mu)} \qquad (3)$

Countercurrent Flow as $\quad \varepsilon = \dfrac{1 - e^{-NTU\,(1-\mu)}}{1 - \mu e^{-NTU\,(1-\mu)}} \qquad (4)$

where

$$NTU = \frac{U.S}{C_{min}} \qquad\qquad from\,(3.19)$$

$$Q_{actual} = 0.5 \times 3850 \times ((273+100) - (273+70)), \frac{kg}{s}.\frac{J}{kg.K}.K = 57{,}750W \quad (5)$$

$$T_{c,out} = 303 + \frac{57{,}750}{0.92 \times 4180} = 318K \qquad (6)$$

$$C_h = 0.5 \times 3850\frac{kg}{s}.\frac{J}{kg.K}. = 1925\frac{W}{K} \qquad (7)$$

$$C_c = 0.92 \times 4180\frac{kg}{s}.\frac{J}{kg.K}. = 3845\frac{W}{K} \qquad (8)$$

Hence $C_c > C_h$ and $C_{min} = C_h$. We can, now compute C_{max} as

$$Q_{max} = C_{min}\left(T_{h,in} - T_{c,in}\right) = 1925 \times (373 - 303), \frac{W}{K}.K = 134{,}750W \quad (9)$$

$$NTU = \frac{U.S}{C_{min}} = \frac{500 \times 2.5}{1925}, \frac{W}{m^2.K}.\frac{m^2}{1}.\frac{K}{W} = 0.65 \qquad (10)$$

$$\mu = \frac{C_{min}}{C_{max}} = \frac{1905}{3845}, \frac{W}{.K}.\frac{K}{W} = 0.50 \qquad (11)$$

The fluid flow directions and terminal temperatures for co-current and counter-current flow are as illustrated below.

For counter-current flow

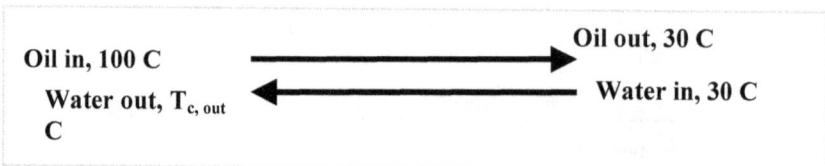

Oil in, 100 C — Oil out, 30 C
Water out, $T_{c,out}$ C — Water in, 30 C

From basic definitions

$$\varepsilon = \frac{Q_{actual}}{Q_{max}} = \frac{57750}{134750} = 0.4286 \qquad (12)$$

$$\varepsilon = \frac{C_c}{C_{min}} \cdot \frac{(T_{c,out} - T_{c,in})}{(T_{h,in} - T_{c,in})} = \frac{3845}{1925} x \frac{(318 - 303)}{(373 - 303)} = 0.4280 \qquad (13)$$

From the derived expressions, equation (4)

$$\varepsilon = \frac{1 - e^{-NTU(1-\mu)}}{1 - \mu e^{-NTU(1-\mu)}} = \frac{1 - e^{-0.65(1-0.5)}}{1 - 0.5 e^{-0.65(1-0.5)}} = \frac{0.2775}{0.6387} = 0.4348 \quad (14)$$

For co-current flow

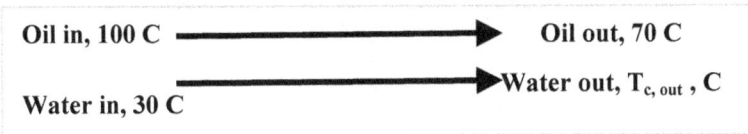

Oil in, 100 C   ━━━━━━━━━━━━▶   **Oil out, 70 C**

━━━━━━━━━━━━▶ **Water out, T$_{c,\,out}$, C**

Water in, 30 C

$$\varepsilon = \frac{Q_{actual}}{Q_{max}} = \frac{57750}{134750} = 0.4286 \qquad (15)$$

$$\varepsilon = \frac{C_c}{C_{min}} \cdot \frac{(T_{c,out} - T_{c,in})}{(T_{h,in} - T_{c,in})} = \frac{3845}{1925} x \frac{(318 - 303)}{(373 - 303)} = 0.4280 \qquad (16)$$

From the derived expressions, equation (3)

$$\varepsilon = \frac{1 - e^{-NTU(1+\mu)}}{(1+\mu)} = \frac{1 - e^{-0.65(1+0.5)}}{1 + 0.5} = \frac{0.6228}{1.5} = 0.4152 \qquad (17)$$

Equations (12) to (17) show that, from the basic definitions of effectiveness, there is no difference between co-current and counter-current flow. The derived expressions, however, yield results which are not only different from those obtained from the basic definitions but also are different for co-current and counter-current flow. This may be the result of simplifying assumptions used in their derivation but the results from the basic definitions should be more reliable because they are based on thermodynamic considerations.

Example 3.9

Water vapour, from an evaporator, is condensed at 49 C in the shell of a single pass exchanger containing 320 mild steel tubes, 25 mm ID x 31 mm OD and 3.66 m. long. The cooling water, passing through the tubes, is available at 21 C and its pressure drop is 2068.5 Pa.

Calculate the cooling water consumption and obtain an estimate of the water outlet temperature using Reynold's analogy. The resistance to heat transfer on the steam side and through the tube wall may be neglected. (For water: specific heat capacity = 4.187 kJ/kgK; viscosity = 0.0008599 Ns/m^2; density = 998 kg/m^3. The relative roughness of mild steel pipe is $\varepsilon/D = 0.001$)

Answer

Because the thermal data, given for this problem, are insufficient to solve the problem, we have an opportunity to demonstrate another way in which fluid friction data can be used. The Reynolds analogy assumes that there is, virtually, no boundary layer and that Pr = 1. Consequently,

$$St = \frac{h_i}{\rho C p u} = \frac{\tau}{\rho u^2} \qquad (1)$$

Equation (1) gives us a value of the film heat transfer coefficient, h, with which we can make a thermal energy balance. Thus, for the n tubes

$$n.h_i.\pi.D_i.L.\Delta T_{lm} = n.\pi.\frac{D_i^2}{4}.\rho.u.Cp(T_{w,out} - T_{w,in}) \quad or$$

$$\frac{h_i}{\rho C p u} = \frac{D_i.(T_{w,out} - T_{w,in})}{4.L.\Delta T_{lm}} \qquad (2)$$

where

$$\Delta T_{lm} = \frac{(T_{shell} - T_{w,in}) - (T_{shell} - T_{w,out})}{\ln\left(\dfrac{T_{shell} - T_{w,in}}{T_{shell} - T_{w,out}}\right)} \qquad (3)$$

From equations (1), (2) and (3)

$$\frac{\tau}{\rho u^2} = \frac{D_i.(T_{w,out} - T_{w,in})}{4.L.\Delta T_{lm}}$$

$$= \frac{D_i.(T_{w,out} - T_{w,in})}{4.L.} x \frac{\ln\left(\dfrac{T_{shell} - T_{w,in}}{T_{shell} - T_{w,out}}\right)}{(T_{shell} - T_{w,in}) - (T_{shell} - T_{w,out})}$$

That is

$$\ln\left(\frac{T_{shell} - T_{w,in}}{T_{shell} - T_{w,out}}\right) = \frac{4.L}{D_i} x \frac{\tau}{\rho u^2} \qquad (4)$$

τ is the shear stress on the inside surface of the tube and L is the length of the tubes. The pressure drop inside tubes, for a single shell, single tube pass heat exchanger, allowing for entrance and exit losses, is given from equation (3.43) by

$$\frac{\Delta P}{\rho u^2} = 4.\frac{L}{D_i}.\frac{\tau}{\rho u^2} + 1 \qquad (5)$$

since $K_{exit} = K_{entrance} = \rho u^2$.
But $Re = \rho D_i u/\mu$ and $u = \mu Re/\rho D_i$ so that

$$\frac{\Delta P}{\rho u^2} = \frac{\Delta P}{\rho.\frac{\mu^2 Re^2}{\rho^2 D_i^2}} = 4.\frac{L}{D_i}.\frac{\tau}{\rho u^2} + 1$$

$$or \quad Re = \sqrt{\frac{\rho D_i^2 \Delta P}{\mu^2\left(4.\frac{L}{D_i}.\frac{\tau}{\rho u^2} + 1\right)}} \qquad (6)$$

From the Colebrook equation

$$\frac{1}{\sqrt{f}} = 2.457\ln\left(\frac{1}{\frac{0.888}{Re\sqrt{f}} + 0.27\frac{\varepsilon}{D}}\right) \qquad from\ (1.64)$$

where $f = \frac{\tau}{\rho u^2}$.

Substituting the given values in equations (6) we get

$$Re = \sqrt{\frac{998 x 0.025^2 x 20685}{(0..0008599^2 x \left(4x\frac{3.66}{0.025} x f + 1\right)}} = \frac{41,771.98}{\sqrt{585.6 f + 1}} \qquad (7)$$

Similarly for equation (1.64)

$$\frac{1}{\sqrt{f}} = 2.457\ln\left(\frac{1}{\frac{0.888}{Re\sqrt{f}} + 0.27 x 0.001}\right) = 2.457\ln\left(\frac{1}{\frac{0.888}{Re\sqrt{f}} + 0.00027}\right) \qquad (8)$$

Equation (8) can be re-arranged as

$$Re = \frac{0.888}{\sqrt{f}\left(e^{-\dfrac{1}{2.457\sqrt{f}}} - 0.00027\right)} \qquad (9)$$

Equations (6) and (9) can be solved by either a graphical or trial by error procedure. Using a graphical procedure, we get the table below which lists the values of Re calculated from equations (6) and (9).

f	Re(6)	Re(9)
0.001	33173.27	-105004
0.0015	30478.32	-94469.9
0.002	28348.83	-125336
0.0025	26611.22	820898.7
0.003	25158.37	50236.24
0.0035	23920.13	19786.53
0.004	22848.4	10524.95
0.0045	21908.92	6464.756

When these are plotted we get the graph below which shows two points of intersection at $f = 0.0021$ and $f = 0.00337$.

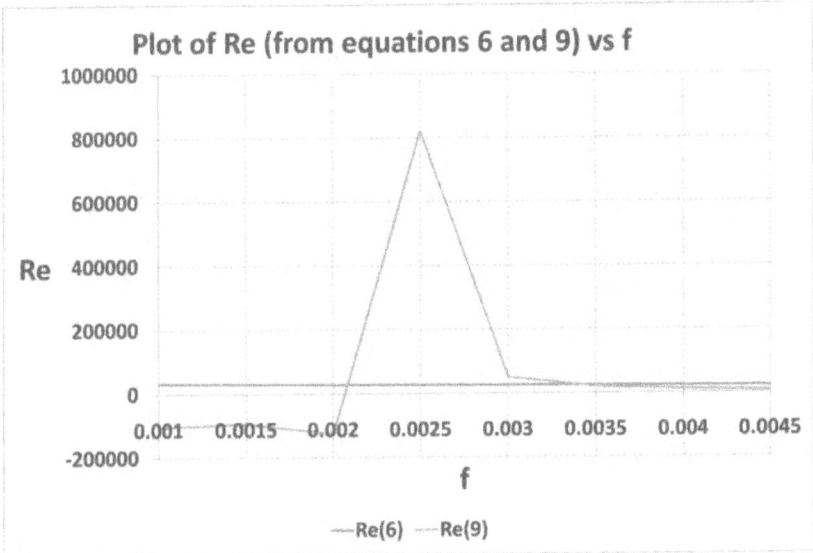

Plot of Re (from equations 6 and 9) vs f

These give values of Re (6) of 27974.1 and Re (9) of -147,852 for f = 0.0021 and Re (6) of 24,224 and Re (9) of 24,208 for f = 0.00337. The acceptable solution is f = 0.00337 for which Re (6) and Re (9) differ by only 0.07%.

From $u = \mu\, Re/\rho D_i$, $\quad u = \dfrac{0.000859 \times 24,224}{998 \times 0.025}, \dfrac{kg}{m.s}.\dfrac{m^3}{kg}.\dfrac{1}{m} = 0.835\dfrac{m}{s}$

Coolingwater flowrate

$= n.\pi.\dfrac{D_i^2}{4}.\rho.u = 320x\,\pi\,x\dfrac{0.025^2}{4}x998x0.835,\dfrac{m^2}{1}.\dfrac{kg}{m^3}.\dfrac{m}{s} = 130.9\dfrac{kg}{s}$ *Ans.*

To estimate the outlet water temperature, we use equation (4), with f = 0.00337. That is

$\ln\left(\dfrac{T_{shell} - T_{w,in}}{T_{shell} - T_{w,out}}\right) = \dfrac{4.L}{D_i}x\dfrac{\tau}{\rho u^2} = \dfrac{4\,x\,3.66}{0.025}x0.00337 = 1.9735$

$\dfrac{322 - T_{w,out}}{322 - 294} = e^{-1.9735}$ and $T_{w,out} = 322 - 28x0.1390 = 3181\,K\,(45.1C)\,Ans.$

Example 3.10

Water vapour, from an evaporator, is condensed, at 49 C, in the shell of an exchanger containing four tube passes, each of length, 2.44m. Each pass consists of 110 tubes, 25.4 mm ID x 31.8 mm OD. The cooling water, passing through the tubes, is available at 21 C and the allowable pressure drop across the exchanger is 8733 Pa. The resistances of the tube wall and on the steam side may be neglected.

Calculate the cooling water consumption and the corresponding condensation rate using a) Reynold's analogy, b) Prandtl's analogy and c) Colburn's analogy.

For water, mean density is 998 kg/m^3, mean viscosity is 0.00086 Ns/m^2 and the mean thermal conductivity is 0.609 W/mK. (Sargent, 1966).

Answer

The Reynolds analogy assumes that there is, virtually, no boundary layer and that Pr = 1. Consequently,

$$St = \frac{h_i}{\rho C p u} = \frac{\tau}{\rho u^2} \tag{1}$$

The Prandtl (Taylor-Prandtl) analogy states that, in practice, there is a laminar boundary layer as well as a turbulent region of flow, in any conduit, and that each is associated with clearly defined velocities, temperatures, etc. Consequently, for $0.5 < \text{Pr} < 30$,

$$St = \frac{h_i}{\rho C p u} = \frac{\tau}{\rho u^2} \left[\frac{1}{1 + 1.15(\text{Pr} - 1)\text{Re}^{-\frac{1}{8}}} \right] \tag{2}$$

The Colburn analogy accepts the von Karman proposition that the transition from laminar to turbulent flow is not sharp but occurs in a smooth transition from a laminar sub-layer through a buffer layer through a turbulent boundary layer to fully developed turbulent flow. It is, however, concerned only with the turbulent boundary layer. It states that, for such a layer and for $0.5 < \text{Pr} < 50$,

$$St = \frac{h_i}{\rho C p u} = \frac{\tau}{\rho u^2} . \text{Pr}^{-\frac{2}{3}} \tag{3}$$

Recall that the pressure drop, ΔP, is given, from (3.43), by

$$\frac{\Delta P}{\rho u^2} = 4 . \frac{L}{D_i} . \frac{\tau}{\rho u^2} \tag{4}$$

when end effects may be neglected. Since $u = \mu \, \text{Re}/\rho D_i$, equation (4) becomes

$$\frac{\tau}{\rho u^2} = \frac{\rho D_i^3}{4 \mu^2 \text{Re}^2} \frac{\Delta P}{L} \tag{5}$$

Recall, also, that the Moody chart equation for smooth pipes is given as

$$\frac{1}{\sqrt{f}} = 2.457 \ln\left(\frac{\text{Re}\sqrt{f}}{0.888} \right) \tag{6}$$

where f is the friction factor, related to the Fanning friction factor, C_f, as

$$f = \frac{C_f}{2} = \frac{\tau}{\rho u^2} \tag{7}$$

From (5) and (7)

$$f = \frac{\rho D_i^3}{4 \mu^2 \text{Re}^2} \frac{\Delta P}{L} \tag{8}$$

Substituting the given values into equation (8) and rearranging

$$Re\sqrt{f} = \frac{998x(0.0254)^3}{4x(0.00086)^2} x \frac{8733}{4x2.44}, \frac{kg}{m^3} \cdot \frac{m^3}{1} \cdot \frac{m^2.s^2}{kg^2} \cdot \frac{1}{m} \cdot \frac{kg.m}{s^2.m^2} = 22241 \,(9)$$

From (6) and (9)

$$\frac{1}{\sqrt{f}} = 2.457\ln\left(\frac{Re\sqrt{f}}{0.888}\right) = 2.457\ln\left(\frac{22241}{0.888}\right) = 19.2282 \quad or \quad f = 0.0027 (10)$$

From (9) and (10)

$$Re = 22241 \, x \, 19.2282 = 42,765.4 \qquad (11)$$

The mass flowrate, M, is then

$$M = \rho u \pi \frac{D_i^2}{4} = \mu.\pi.n.Re.\frac{D_i}{4}$$

$$= 0.00086x \, \pi \, x110x \, 42,765.4 \, x \frac{0.0254}{4}, \frac{kg}{m.s} \cdot \frac{m}{1} = 80.71\frac{kg}{s} \quad Ans.$$

where n is the number of tubes.

To calculate the condensation rate using the different analogies between heat and momentum transfer

Let W be the condensation rate and λ, the latent heat of condensation. By a heat balance

$$W.\lambda = \rho u n \pi \frac{D_i^2}{4} Cp \left(T_{w,out} - T_{w,in}\right) = h_0 \, \pi \, D_0 \, n \, L \, \Delta T_{lm} \qquad (12)$$

where ΔT_{lm} is the log mean temperature difference given by

$$\Delta T_{lm} = \frac{(T_{shell} - T_{w,in}) - (T_{shell} - T_{w,out})}{\ln\left(\dfrac{T_{shell} - T_{w,in}}{T_{shell} - T_{w,out}}\right)} \qquad (13)$$

From (12) and (13)

$$\frac{h_0}{\rho u Cp} = St = \frac{D_i^2}{4 D_0 L}.\ln\left(\frac{T_{shell} - T_{w,in}}{T_{shell} - T_{w,out}}\right) \qquad (14)$$

For water, $\lambda = 2385$ kJ/kg and Cp = 4.187 kJ/kg.K.

$$Pr = \frac{Cp \, \mu}{k} = \frac{4187x0.00086}{0.609}, \frac{J}{kg.K} \cdot \frac{kg}{m.s} \cdot \frac{s.m.K}{J} = 5.92 > 1 \quad (15)$$

This value of Pr shows, *a priori*, that the Reynolds analogy is not, strictly, applicable to this situation. Nevertheless, for comparison, we shall apply the Reynolds analogy to the problem as follows below.

a.   Reynolds Analogy

From (1), (14) and (10)

$$St = \frac{\tau}{\rho u^2} = f = \frac{D_i^2}{4\,D_0\,L}.\ln\left(\frac{T_{shell} - T_{w,in}}{T_{shell} - T_{w,out}}\right) = 0.0027 \quad (16)$$

Substituting given values

$$\ln\left(\frac{322 - 294}{322 - T_{w,out}}\right) = \frac{4\,x\,0.0318\,x\,4\,x\,2.44}{(0.0254)^2}\,x\,0.0027 = 5.1956 \quad (17)$$

From which $T_{w,\,out}$ = 321.85 K (48.85 C). The condensation rate is, therefore, from equation (12)

$$W = 80.71\,x\,4187\,x\frac{(321.85 - 294)}{2,385,000},\frac{kg}{s}.\frac{J}{kg.K}.\frac{K}{1}.\frac{kg}{J} = 3.95\frac{kg}{s} \quad (18)$$

b.   Prandtl Analogy

From (2),

$$St = f.\left[\frac{1}{1 + 1.15(Pr - 1)Re^{-\frac{1}{8}}}\right] = 0.0027x\left[\frac{1}{1 + 1.15(5.92 - 1)\,x\,(427654)^{-0.125}}\right]$$

$$= 0.001084 \quad (19)$$

From (19) and (14)

$$St = 0.001084 = \frac{D_i^2}{4\,D_0\,L}.\ln\left(\frac{T_{shell} - T_{w,in}}{T_{shell} - T_{w,out}}\right) \quad (20)$$

Substituting given values

$$\ln\left(\frac{322 - 294}{322 - T_{w,out}}\right) = \frac{4\,x\,0.0318\,x\,4\,x\,2.44}{(0.0254)^2}\,x\,0.001084 = 2.0859 \quad (21)$$

From which $T_{w,\,out}$ = 318.5 K (45.5 C). The condensation rate is, therefore, from equation (12)

$$W = 80.71\,x\,4187\,x\frac{(3185 - 294)}{2,385,000},\frac{kg}{s}.\frac{J}{kg.K}.\frac{K}{1}.\frac{kg}{J} = 3.47\frac{kg}{s} \quad (22)$$

c.   Colburn's Analogy

From (3),

$$St = \frac{h_i}{\rho Cpu} = \frac{\tau}{\rho u^2}.Pr^{-\frac{2}{3}} = f Pr^{-\frac{2}{3}} = 0.0027x(5.92)^{-\frac{2}{3}} = 0.000825 \quad (23)$$

From (23) and (14)

$$St = 0.000825 = \frac{D_i^2}{4 D_0\ L}.\ln\left(\frac{T_{shell} - T_{w,in}}{T_{shell} - T_{w,out}}\right) \quad (24)$$

Substituting given values

$$\ln\left(\frac{322-294}{322-T_{w,out}}\right) = \frac{4 x 0.0318 x 4 x 2.44}{(0.0254)^2} x 0.000825 = 1.5875 \quad (25)$$

From which $T_{w,\ out}$ = 316.3 K (43.3 C). The condensation rate is, therefore, from equation (12)

$$W = 80.71x4187x\frac{(316.3-294)}{2,385,000}\ \frac{kg}{s}.\frac{J}{kg.K}.\frac{K}{1}\frac{kg}{J} = 3.16\frac{kg}{s} \quad (26)$$

These results may be summarised in the table below

	Reynolds Analogy	Prandtl Analogy	Colburn Analogy
Water exit temperature, K	321.85	318.5	316.3
Condensation rate, kg/s	3.95	3.47	3.16

Except for the Reynolds analogy which is inapplicable to this problem, the other two analogies give comparable results although the Colburn analogy is less tedious to use.

Example 3.11

15.13 kg/s of brine are to be cooled from +5 C to -15 C in a shell and tube heat exchanger, using ammonia evaporating at -20 C in the shell. The allowable pressure drop for the brine is 0.31 bar.

Calculate the number and length of tubes required using mild steel tubes of 12.5 mm OD x 16 SWG.

For brine, specific gravity = 1.23; specific heat = 2.763 kJ/kg.K; viscosity = 0.0024 Ns/m²; thermal conductivity = 0.554 W/m.K. The thermal conductivity of the steel = 45 W/m.K. The heat transfer coefficient for boiling ammonia = 1590 W/m².K. (Sargent, 1966)

Answer

Let U_i be the overall heat transfer coefficient based on the inside tube surface, S_i the total inside tube surface area and ΔT_{lm} the log mean temperature difference. By a heat balance

$$\rho u n n \pi \frac{D_i^2}{4} Cp \left(T_{b,in} - T_{b,out}\right) = U_i \, N n \pi D_i \, L \Delta T_{lm} \qquad (1)$$

where

$\quad u =$ velocity of fluid in each tube
$\quad n =$ number of tubes per pass
$\quad T_{b,\,in} =$ inlet temperature of the brine inside the tubes
$\quad T_{b,\,out} =$ outlet temperature of the brine inside the tubes
$\quad L =$ length of tubes per pass
$\quad N =$ number of tube passes, assumed $= 1$

Thus

$$\frac{4L}{D_i} = \frac{\rho u \, Cp \left(T_{b,in} - T_{b,out}\right)}{U_i \, \Delta T_{lm}} \qquad (2)$$

Since the pressure drop, ΔP, is given, from (3.43), by

$$\frac{\Delta P}{\rho u^2} = 4 . \frac{L}{D_i} . \frac{\tau}{\rho u^2} \qquad \text{from (3.43)}$$

when end effects may be neglected, we get from (2) and (3.43)

$$\frac{\Delta P}{\rho u^2} = \frac{\rho u \, Cp \left(T_{b,in} - T_{b,out}\right)}{U_i \, \Delta T_{lm}} . \frac{\tau}{\rho u^2} \qquad (3)$$

which can be re-arranged to

$$u^3 = \frac{\Delta P U_i \, \Delta T_{lm}}{\rho^2 \, Cp \left(T_{b,in} - T_{b,out}\right) f} \qquad (4)$$

From the data given:

$\qquad \Delta P = \quad 0.31 \times 1.013 \times 10^5 = 31403 \text{ N/m}^2$
$\qquad \rho = \quad 1.23 \times 998 = 1227.54 \text{ kg/m}^3$
$\qquad Cp = \quad 2{,}763 \text{ J/kg.K}$
$\qquad \Delta T_{lm} = \quad \dfrac{(278-253)-(258-253)}{\ln\dfrac{(278-253)}{(258-253)}} = 12.43 K$

$\qquad T_{b,\,in} - T_{b,\,out} = \quad 278 - 258 = 20 \text{ K}$

Substituting all these in (4)

$$u^3 = \frac{31403x\,U_i\,x12.43}{(12275.4)^2\,x2763x\,20x\,f}\,,\frac{kg.m}{s^2.m^2}\cdot\frac{J}{s.m^2.K}\cdot\frac{K}{1}\cdot\frac{m^6}{kg^2}\cdot\frac{kg.K}{J}\cdot\frac{1}{K}$$

$$= 4.6877x10^{-6}\,\frac{U_i}{f}\,,\frac{m^3}{s^3} \qquad\qquad (5)$$

Also

$$\frac{1}{U_i} = \left[\frac{1}{h_i} + \frac{D_i}{2k}\ln\left(\frac{D_0}{D_i}\right) + \frac{D_i}{h_0\,D_0}\right] \qquad from\ (2.20)$$

$$= \frac{1}{h_i} + \frac{0.0095}{2\,x45}\ln\left(\frac{0.0125}{0.0095}\right) + \frac{0.0095}{1590x0.0125} = \frac{1}{h_i} + 5.0696x10^{-4} \quad (6)$$

where, for 12.5 mm OD tubes, D_i = 9.5 mm, D_0 = 12.5 mm.

To evaluate h_i, we shall use the Dittus - Boelter equation, given as

$$Nu_i = \frac{h_i\,D_i}{k} = 0.023Re^{0.8}\,Pr^{0.4} \qquad\qquad from\ (2.28)$$

But

$$Pr = \frac{Cp\,\mu}{k} = \frac{2763x0.0024}{0.554}\,,\frac{J}{kg.K}\cdot\frac{kg}{m.s}\cdot\frac{s.m.K}{J} = 11.97 \qquad (7)$$

Hence, from (7) and (2.28)

$$h_i = \frac{0.554}{0.0095}x0.023Re^{0.8}\,x11.97^{0.4} = 3.62Re^{0.8} \qquad (8)$$

Substitution of (8) in (6) results in the so called Wilson's equation.

f may be obtained from Moody charts or from the equation on which the chart is based. Let us try the Churchill equations (1.75), (1.75a) and (1.75b), which, though complicated, represent an improvement on the Moody chart equations. Thus

$$f = \left[\left(\frac{8}{Re}\right)^{12} + \frac{1}{(A+B)^{3\!/\!2}}\right]^{1\!/\!12} \qquad\qquad (1.75)$$

where

$$A = \left[2.457\ln\left(\frac{1}{\left(\frac{7}{Re}\right)^{0.9} + 0.27\frac{\varepsilon}{D}}\right)\right]^{16} \qquad (1.75a)$$

and

153

$$B = \left(\frac{37,530}{Re}\right)^{16} \qquad (1.75b)$$

$\frac{\varepsilon}{D}$ is the relative roughness of the pipe, equal to zero for a smooth pipe.

For any value of Re, therefore, h_i can be obtained from (8), which is then used to calculate U_i in (6). This Re, however, must be the Re that corresponds to the u and f in equation (5). This means a trial and error or graphical procedure. For a trial and error procedure, for example, guess Re = 5000. From (8),

$$h_i = 3.62 x 5000^{0.8} = 3295\frac{W}{m^2 .K} \qquad (9)$$

From (6),

$$\frac{1}{U_i} = \frac{1}{3295} + 5.0696x10^{-4} \quad or \quad U_i = 12339\frac{W}{m^2 .K} \quad (10)$$

From (1.75b), B = 0. From (1.75a)

$$A = \left[2.457\ln\left(\frac{1}{\left(\frac{7}{5000}\right)^{0.9}}\right)\right]^{16} = 3.9515x10^{18} \qquad (11)$$

From (1.75)

$$f = \left[\left(\frac{8}{Re}\right)^{12} + \frac{1}{(A+B)^{3\!/\!2}}\right]^{1\!/\!12}$$

$$= \left[\left(\frac{8}{5000}\right)^{12} + \frac{1}{\left(3.9515x10^{18}\right)^{3\!/\!2}}\right]^{1\!/\!12} = 0.004736 \qquad (12)$$

From (5), (10) and (12)

$$u^3 = 4.6877x10^{-6}\frac{U_i}{f}, \frac{m^3}{s^3} = 4.6877x10^{-6} x \frac{12339}{0.004736}. \frac{m^3}{s^3} \quad or$$

$$u = 1.07\frac{m}{s} \qquad (13)$$

To check that the guessed Re is the right one, recalculate Re as

$$Re = \frac{\rho D_i u}{\mu} = \frac{1227.54 x 0.0095 x 1.07}{0.0024}, \frac{kg}{m^3}.\frac{m}{1}.\frac{m}{s}.\frac{m.s}{kg} = 51991$$

This is 3.98% higher than the guessed value and is usually an

154

acceptable margin of error in trial and error procedures but we can improve on it. If we guess a new Re = 5199, we shall get, as before, from equation (8), h_i = 3399.7 W/m²K; from equation (6), U_i = 1248.3 W/m²K; from equation (1.75a), A = 4.3442 x 10^{18}; from equation (1.75), f = 0.00468; from equation (5), u = 1.0773m/s to give a Re = 5235 (in error over Re = 5199 by 0.7%). This level of error is more acceptable but may not be worth the extra effort in certain situations.

To calculate the length of tubes, we use equation (2).

$$L = \frac{D_i}{4} \cdot \frac{\rho u C_p \left(T_{b,in} - T_{b,out}\right)}{U_i \Delta T_{lm}}$$

$$= \frac{0.0095}{4} x \frac{12275.4 x 1.0773 x 2763 x 20}{12483 x 12.43} \frac{m}{1} \cdot \frac{kg}{m^3} \cdot \frac{m}{s} \cdot \frac{J}{kg.K} \cdot \frac{K}{1} \cdot \frac{s.m^2.K}{J} \cdot \frac{1}{K}$$

$$= 11.186m. \quad Ans$$

To calculate the number of tubes per pass, n,

$$\rho u n \pi \frac{D_i^2}{4} = 15.13 \frac{kg}{s} \quad or$$

$$n = \frac{4 x 15.13}{12275.4 x 1.0773 x \pi x 0.0095^2}, \frac{kg}{s} \cdot \frac{m^3}{kg} \cdot \frac{s}{m} \cdot \frac{1}{m^2}$$

$$= 161.4 \cong 162 \; tubes \, per \, pass \quad Ans.$$

Example 3.12

Saturated steam, at 3.45 x 10^4 N/m², guage pressure, is condensed in the shell of a shell and tube heat exchanger and the condensate leaves at its boiling point. Cooling water is available at 15.6 C and when its flowrate is 227.3 gal/ min, the steam condensation rate is 1.58 kg/s. What will be the condensation rate when the water flowrate is increased to 341 gal/ min?

Assume that the heat transfer coefficient between the cooling water and the tube surface is proportional to the 0.8 power of the Reynold's number. Neglect all other heat transfer resistances.

Answer

The effectiveness - NTU relationship, for .a co-current exchanger, is given by

$$\varepsilon = \frac{Q_{actual}}{Q_{max}} = \frac{1 - e^{-NTU\,(1+\mu)}}{1+\mu} \qquad (1)$$

where $\mu = C_{min}/C_{max}$. If one temperature is constant, it is equivalent to having $C_h = \infty$ for the fluid in question. Thus $\mu = 0$ and equation (1) reduces to

$$\varepsilon = \frac{Q_{actual}}{Q_{max}} = 1 - e^{-NTU} \quad or \quad NTU = \ln\left(\frac{1}{1-\varepsilon}\right) \qquad (2)$$

For steam at 34500 N/m² guage pressure, (135850, N/m² absolute pressure), the saturation temperature is 108.34 C and latent heat is 2234.7 kJ/kg. At the cooling water flowrate of 227.3 gal/ min. (1 gal = 4 litres), heat energy released by steam condensation, Q, is

$$Q = 1.58 \times 22347, \frac{kg}{s} \cdot \frac{kJ}{kg} = 35308\,kW \qquad (3)$$

The cooling water exit temperature, $T_{w,\,exit}$ is, therefore,

$$T_{w,exit} = T_{w,in} + \frac{Q}{m_w \cdot Cp_w}$$

$$= 288.6\,K + \frac{35308 \times 60 \times 1000}{227.34 \times 4 \times 998 \times 4.187}, \frac{kJ}{s}\frac{min}{gal}\frac{s}{min}\frac{gal}{litres}\frac{litres}{m^3}\frac{m^3}{kg}\frac{kg.K}{kJ}$$

$$= 344.4\,K\,(71.4\,C) \qquad (4)$$

From (3.17),

$$\varepsilon = \frac{Q_{actual}}{Q_{max}} = \frac{C_h\left(\theta_{h,in} - \theta_{h,out}\right)}{C_{min}\left(\theta_{h,in} - \theta_{c,in}\right)} = \frac{C_c\left(\theta_{c,out} - \theta_{c,in}\right)}{C_{min}\left(\theta_{h,in} - \theta_{c,in}\right)} \qquad (3.17)$$

Since $C_h \to \infty$, $C_c = C_{min}$, and

$$\varepsilon = \frac{Q_{actual}}{Q_{max}} = \frac{C_c\left(T_{w,out} - T_{w,in}\right)}{C_{min}\left(T_{h,in} - T_{w,in}\right)} = \frac{\left(T_{w,out} - T_{w,in}\right)}{\left(T_{h,in} - T_{w,in}\right)} \qquad (5)$$

Hence, from (4) and (5)

$$\varepsilon = \frac{\left(T_{w,out} - T_{w,in}\right)}{\left(T_{h,in} - T_{w,in}\right)} = \frac{(344.4 - 288.6)}{(381.34 - 288.6)} = 0.60 \qquad (6)$$

From (2)

$$NTU = \ln\left(\frac{1}{1-\varepsilon}\right) = \ln\left(\frac{1}{1-0.6}\right) = 0.9163 \qquad (7)$$

Since $u \propto Re^{0.8} \propto m^{0.8}$ then $NTU = \frac{UC}{mCp} \propto m^{0.2}$. That is

$$\frac{NTU_{227}}{NTU_{341}} = \left(\frac{m_{w,341}}{m_{w,227}}\right)^{0.2} \tag{8}$$

Thus, at a cooling water flowrate of 341 gal/min

$$NTU_{341} = NTU_{227} \, x \left(\frac{m_{w,227}}{m_{w,341}}\right)^{0.2} = 0.9163x \left(\frac{227.3}{341}\right)^{0.2} = 0.8449 \tag{9}$$

From (2)

$$\varepsilon = 1 - e^{-NTU} = 1 - e^{-0.8449} = 0.57 \tag{10}$$

From (1)

$$Q_{actual} = \varepsilon.Q_{max} = 0.57x \frac{341x4x998}{60x1000} x \, 4.187(381.4 - 288.6) = 50248\,kW \tag{11}$$

$$\frac{gal}{min} \frac{min}{s} \frac{litres}{gal} \frac{m^3}{litres} \frac{kg}{m^3}.\frac{kJ}{kg.K} K$$

$$\text{Condensation rate} = \frac{Q_{actual}}{latent\,heat} = \frac{50248}{22345}, \frac{kJ}{s}.\frac{kg}{kJ} = 2.25 \frac{kg}{s} \quad Ans.$$

References for Chapter Three

1 Coulson J. M, Richardson J. F., and Sinnott R. K., Chemical Engineering Vol. 6; Design; Pergamon Press, Oxford. U.K., 1983.

2 Kays W M. and London A. I, Compact Heat Exchangers; 2nd Edition, McGraw-Hill Book Company; New York. 1964

3 Peters M. S. and Timmerhaus K. D., Plant Design and Economics for Chemical Engineers; 3rd Edition; Int'l Student Edition; McGraw-Hill International Book Company, London, 1985.

4 Sargent R. W. II., Class Notes in Heat Transfer, Imperial College, London, 1965/66

CHAPTER FOUR
THE OPTIMISATION OF HEAT EXCHANGERS

The optimisation of heat exchangers, in commercial practice, like that of other components of a complex manufacturing plant, is, usually, done at three levels. At the primary level, such as the design, manufacturing or selection of an individual unit, optimisation is defined in respect of the estimated initial purchase and estimated operating costs of the heat exchanger. At this level may, also, be considered small scale and pilot plants in which excursions from optimum can have serious consequences to efficiency, profitability or reliability of design and operating data.

At the secondary level where, not one but, a number of exchangers are connected in series, parallel or in any other combinations in the so called heat exchanger networks (HENs), the optimisation problem becomes a process optimisation problem in which the whole heating and cooling duty in the plant, or of a system of heat exchangers, is optimised.

At the tertiary level (the plant optimisation level), when the heat exchanger is seen, not only as an individual unit or part of a heat exchanger network but also, as a part of a complex manufacturing/processing plant with its control and operating costs and problems, the optimisation is considered, additionally, in terms of controllability and overall plant/process profitability *viz-a-viz* other expensive, unstable, or other key components and systems in the plant. What is optimum in the plant may or may not correspond to the individualised optimum for the heat exchanger or heat exchanger network and vice versa.

We shall be concerned, here, only with introductory treatments of heat exchanger optimisation at the primary and secondary levels while the plant optimisation problem is considered to be outside the scope of this book. Even then, only generalised descriptions will be presented. There have been very rapid advances in optimisation of several aspects of manufacturing processes at each of the above mentioned levels. Some of these are marketed in the form of proprietary software some of which are listed in Table 4.1.

4.1: Optimisation of Heat Exchanger Units

The two most important parameters in the optimisation of heat exchanger units are the total heat transfer area and the pressure drop in the exchanger. This is because the total heat transfer area determines both the initial purchase cost and operating capacity while the pressure drop determines the operating cost of the exchanger in service.

For a given heat transfer rate, higher fluid velocities in the exchanger lead to higher heat transfer coefficients and hence smaller surface areas and lower initial purchase costs. Higher fluid velocities, however, are associated with higher pressure drops which mean higher pumping/compressor costs and hence higher operating costs, which can be up to 20 - 50% per annum of the initial purchase cost (Jegede & Polley, 1992).

Thus, the optimum economic design is one which minimises the total cost of the heat exchanger in terms of the initial purchase cost and the operating cost.

4.2: Methods of Optimising Heat Exchanger Designs

Jegede and Polley (1992) identified the significant milestones in heat exchanger optimisation in the works of McAdams (1954), Jenssen (1969), Steinmeyer (1976 and 1982) and Peters and Timmerhaus (1981). These are summarised below.

4.2.1: Method of McAdams

This was the earliest method to, not only recognise the need for, but also, develop a procedure for, optimising a heat exchanger design. A tubular heat exchanger was considered. The mass velocities of the inside and outside tube fluids were optimised on the basis of the cost of power and the fixed cost of the exchanger. A major weakness of the method, however, was the assumption that the tube and shell sides of the exchanger could be treated independently of each other and had no interaction between them.

4.2.2: Jenssen's Method

This method was developed as a result of the need to be able to estimate the economic power consumption in plate heat exchangers. By defining a *j* parameter, as the specific pressure drop per unit heat transfer, Jenssen (1969) produced graphs which showed the economic optimum, on the assumption that streams on either side of the exchanger have the same flowrates and the same physical properties. This assumption is acceptable for plate, but not for the generality of other types of, heat exchangers.

4.2.3: Steinmeyer's Method

Steinmeyer (1976 & 1982) applied Jenssen's method to shell and tube heat exchangers and developed separate relationships for the shell and tube side heat transfer. Interactions between streams on these sides of the heat exchanger were, however, ignored.

4.2.4: The Method of Peters and Timmerhaus (1981)

This is a comprehensive optimisation procedure which considered the tube-side and shell-side pressure drops simultaneously with the heat transfer area of the exchanger. It assumes a given allowable pressure drop and tube characteristics to optimise the following components and variables of the heat exchanger

i	the film heat transfer coefficients, h_i and h_o
ii	the overall heat transfer coefficients, U_i and U_0
iii	the temperature difference, ΔT
iv	the heat transfer area, S
v	the process and utility fluid flowrates, G_t, and G_s
vi	the flow inside tubes per pass
vii	the number of tubes
viii	tube length
ix	the number of passes (tube and shell)
x	the pressure drop

Its major limitation is that the assumed allowable pressure drop used is not, necessarily, the optimum pressure drop for the heat exchanger (it may be for the plant at the level 3 optimisation). Another problem with the method is the large number of variables used most of which depend on a rather larger number of other variables, some very difficult to predict, especially, in new situations.

4.2.5: The Method of Jegede and Polley (1992)

This method does not assume an *a priori* allowable pressure drop and is general enough, in the initial formulation of the basic equations, to be applicable to any direct heat exchanger type. According to its authors, it will, still, estimate the optimal film heat transfer coefficients on the tube and shell sides as well as the optimal heat transfer surface when a set allowable pressure drop is given. It will, also, estimate the optimum pressure drop when this is not known. In all cases, the interactions between tube side and shell side pressure drop and the surface area for heat transfer are considered simultaneously.

Only the methods of Peters and Timmerhaus (1981) and Jegede and Polley (1992) will, however, be described. Further information may be obtained from the original sources and their various updates.

4.2.6: Computer Aided Optimisation Methods

Computer aided design is, currently, in vogue and a large number of computer programs are now available for thermal and mechanical design of heat exchangers. These programs provide a practical and powerful tool for design engineers and produce significant savings in time and money (Pase, 1986).

Designing a heat exchanger involves four different areas of expertise: chemical engineering, mechanical engineering, cost estimating and pricing. Pase (1986) estimates that, in manual design, it may take the chemical engineer 2 to 8 hours to do the chemical engineering aspects, the mechanical engineer another 8 to 16 hours, most of it spent in doing ASME and TEMA code calculations, the draftsman a week or two, the cost estimator, who does the detailed cost breakdown of the design, and the sales engineer, who determines mark-up and the ultimate pricing for the job, another 1 to 2 weeks, for a total of three to four weeks to produce a single design of a heat exchanger, which, in the end, may not be the optimum. In contrast, the entire operation, and for an optimal heat exchanger, can be carried out by computer in 20 minutes by a single person.

The two most important features of these computer programs are whether they are integrated or interactive. Integrated programs combine the expertise of many specialised disciplines in one single unified procedure. For example, applied chemistry, thermal engineering, pressure vessel mechanics and cost estimating, in specialised databases,

162

can be called upon at any point in the program, and as many times as required, rather than in a linear fashion as would be the case in manual design, thus leading quickly to optimum designs.

Interactive programs guide the designer through the computer procedure using a logical sequence of questions and answers. This enables the designer to consider a large number of alternatives and the effects of adjusting different variables on the final design and/or cost.

It should be noted that the actual thermal and mechanical design procedures used in the programs may or may not be different from published methods. The advantage is in the speed with which the computer applies the design procedures to a given problem and in the access which can be made to a large database of information. A few examples of available software are listed in the Tables below. These listings are neither endorsements nor recommendations of any software.

Table 4.1: Heat Exchanger Design Software (Lauterbach V., 2010)

Package	Exchanger design	Program description
AC	Air Cooler	Cross flow heat exchangers with bare or finned tubes, includes crimped-finned tubes (spiral wound), partial condensation and additional properties.
BREN	Thermal Oil Heater	Calculation and simulation of thermal oil heaters (with extensive thermal oil data base)
COIL	Coil Type Heat Exchanger	Coil type heat exchangers for gases and liquids
DGK	Pressure Gas Cooler	Partial condensation in multi-stage pressure gas coolers, water as coolant
DNK	Vapour / Post-Vapour Condensation	Calculation of the intermediate temperatures and quantities of vapour / condensate in vapour / post-vapour condensation in heat exchanger groups (12 heat exchanger arrangements)
DP	Double Pipe Heat Exchanger	Heat transfer and pressure loss in double pipe heat exchangers
DPW	Double Pipe Heat Exchanger	Thermal and hydraulic design of a double-pipe heat exchanger
KOAX	Coaxial Heat Exchanger	Heat transfer and pressure loss in coaxial heat exchangers
KOND	Steam	Steam condensers with desuperheating, condensing

	Condenser	and sub-cooling option, TEMA, graphics. Multi-tube heat exchangers
LAM	Plate Fin and Tube	Design of plate fin-and-tube heat exchangers
MESK	Multi-Component Condenser	Calculation of pure substance / multi-component condensers with / without inert gases
MMP	Plate Heat Exchanger	Pressure drop and heat transfer of chevron-type plates, for single-phase media only, no condensation. Incl. essential properties
MUKO	Multi-Component Condenser	Cross flow condensers with bare or finned tubes incl. DDB-Flash (multi-component condensation in the tubes)
RS3	Triple tube	Heat exchanger in concentric triple tube
TANK	Hot storage tank	Heat losses / heating of storage tanks
VERD	Tube bundle evaporator	Vaporization of pure substances at smooth tubes, including tubesheet library TEMA and essential properties.
WAK	Coil type economizer for flue gas	Coil type heat exchangers with up to 14 concentric tube baskets, heating of thermal oil incl. LAMO module
WTS	Shell-and-tube heat exchanger	Basic version of design system for shell-and-tube heat exchangers for single-phase media (liquid / gaseous) with CAD interface, graphical layout of tubesheet, true-to-scale heat exchanger sketch and interface to LV strength calculation programs.

Table 4.2: More Heat Exchanger Design Software

Package	Exchanger design	Vendor
Aerotran	Interactive thermal design - air cooled heat exchangers, etc.	B-JAC Computer Services, Midlothian, Va., USA
B-JAC	Integrated thermal and mechanical design	B-JAC Computer Services, Midlothian, Va., USA
HEXTRAN	Feasibility, Synthesis, Optimisation	Simulation Sciences, Fullerton, Ca., USA
PROCESS	Process Simulation	Simulation Sciences, Fullerton, Ca., USA

4.3: Optimisation of a Heat Exchanger Design by the Method of Peters and Timmerhaus (1981)

This method, with its various simplifications and ramifications, is described in detail in the book by Peters and Timmerhaus (1985). Only the main results and outline procedures are presented here.

The method recognises four costs of importance in the optimisation of heat exchangers, namely, the installed cost, the cost of the utility fluid, the cost of energy required to pump fluid through the tubes, and the cost of energy required to pump fluid through the shell.

It, also, recognises three physical variables as being crucial to an optimum heat exchanger, namely, the heat exchange area, the pressure drop in the tubes and the pressure drop in the shell of the heat exchanger. The pressure drops are treated in terms of the energy required to pump fluids against them. These variables are, then, related to each other as follows:

$$C_T = S_0 \, K_F \, C_{S,0} + F_U \, H_Y \, C_U + S_0 \, E_i \, H_Y \, C_i + S_0 \, E_0 \, H_Y \, C_0 \quad (4.1)$$

where

$C_T =$ total annual variable cost for the heat exchanger and its operation, per annum

$C_{S0} =$ installed cost of the heat exchanger per unit of outside tube heat transfer area, per m^2

$C_U =$ cost of the utility fluid, per kg

$C_i =$ cost of energy required to pump fluid through the tubes, per W

$C_0 =$ cost of energy required to pump fluid through the shell, per W

$S_0 =$ heat transfer area based on outside tube diameter, m^2

$K_F =$ annual fixed charges, including maintenance, expressed as a fraction of initial / installed cost

$F_U =$ flowrate of utility fluid, kg/ h

$H_Y =$ hours of operation per annum

$E_I =$ power loss inside tubes per unit of outside tube area, W/ m^2

$E_0 =$ power loss outside tubes per unit of outside tube area, W/ m^2

Equation (4.1) is general and applicable to any type of heat exchanger. Peters and Timmerhaus (1985), however, used a shell and tube heat exchanger with cross-flow on the shell side, as an example, to optimise equation (4.1) as illustrated below.

A heat energy balance across the exchanger gives

$$Q = F_U \, Cp_U \, (T_{U_2} - T_{U_1}) = F_P \, Cp_P \, (T_{P_2} - T_{P_1}) = U_0 \, S_0 \, \Delta T_{lm} \quad (4.2)$$

where Q, F, Cp, T and U denote the heat duty, flowrate, specific heat capacity, temperature and overall heat transfer coefficient, respectively, and the subscripts u, p, 1 and 2 refer to the utility fluid, the process fluid, entrance and exit conditions, respectively.

From (4.2)

$$F_U = \frac{Q}{Cp_U \, (\Delta T_1 - \Delta T_2 + T_{P_1} - T_{P_2})} \quad (4.3)$$

where $\Delta T_1 = T_{P2} - T_{U1}$ and $\Delta T_2 = T_{P1} - T_{U2}$.

The power loss inside and outside the tubes are calculated as follows.

Power Loss inside Tubes, E_i

$$-\Delta P_i = B_i \cdot \frac{2 f_i \, G_i^2 \, Ln_P}{\rho_i \, D_i \, \phi_i} \quad (4.4)$$

where

$$B_i = 1 + 0.51 \frac{K_i \, n_P \, \Delta T_{lm}}{(T_2 - T_1) . Pr^{\frac{2}{3}}} \cdot \left(\frac{\mu_i}{\mu_w} \right)^{0.28} \quad (4.4a)$$

$$K_1 = \left(1 - \frac{S_i}{S_H} \right)^2 + K_C + 0.5 \left(\frac{n_P - 1}{n_P} \right) \quad (4.4b)$$

$\dfrac{S_i}{S_H} =$ ratio of total inside tube cross-sectional area per pass to header cross-sectional area per pass

$K_C =$ contraction coefficient

$n_P =$ number of tube passes

$f_i =$ Fanning friction factor for isothermal flow based on the arithmetic mean temperature of the fluid $= 0.046\left(\dfrac{\mu_i}{D_i \, G_i} \right)^{0.2}$

$\phi_i =$ correction factor for non-isothermal flow $= 1.02 \left(\dfrac{\mu_i}{\mu_w} \right)^{0.14}$

$$S_i = \frac{\pi D_i^2 \, N_T}{4 n_P} \quad (4.4c)$$

166

$$S_0 = N_T \pi D_0 L \tag{4.5}$$

N_T is the total number of tubes, D_i is the tube inside diameter, G_i is the mass flow velocity, kg/s.m², μ_i is the fluid viscosity evaluated at the arithmetic mean fluid temperature and μ_w is the fluid viscosity evaluated at the average temperature of the inside tube wall surface.

The basic assumptions for the above equations are that there is no change of phase, fluid flow is highly turbulent and the pipes are smooth.

$$E_i = -\frac{\Delta P_i \, G_i \, S_i}{\rho_i \, S_0} = -\frac{\Delta P_i \, G_i \, D_i^2}{4 \rho_i \, D_0 \, L n_P} \tag{4.6}$$

From (4.4) and (4.6) with the value of f_i

$$E_i = 0.023 \cdot \frac{B_i \, \mu_i^{0.2} \, G_i^{2.8} \, D_i^{0.8}}{\rho_i^2 \, D_0^2 \, \phi_i} \tag{4.7}$$

From the Dittus-Boelter equation

$$\frac{h_i \, D_i}{k_i} = 0.023 \left(\frac{G_i \, D_i}{\mu_i} \right)^{0.8} \left(\frac{Cp_i \, \mu_i}{k_i} \right)^{\frac{1}{3}} \left(\frac{\mu_i}{\mu_w} \right)^{0.14} \tag{4.8}$$

From which

$$G_i = \left[\frac{h_i \, D_i^{0.2} \, \mu_i^{0.8}}{0.023 k_i} \cdot \left(\frac{k_i}{Cp_i \, \mu_i} \right)^{\frac{1}{3}} \cdot \left(\frac{\mu_w}{\mu_i} \right)^{0.14} \right]^{1.25} \tag{4.9}$$

Combining (4.7) and (4.9) with the value of ϕ_i

$$E_i = h_i^{3.5} \cdot \frac{12200 B_i \, \mu_i^{1.83} \, D_i^{1.5}}{\rho_i^2 \, D_0^2 \, Cp_i^{1.17} \, k_i^{2.33}} \left(\frac{\mu_w}{\mu_i} \right)^{0.63} = \Psi_i \, h_i^{3.5} \tag{4.10}$$

where

$$\Psi_i = \frac{12200 B_i \, \mu_i^{1.83} \, D_i^{1.5}}{\rho_i^2 \, D_0^2 \, Cp_i^{1.17} \, k_i^{2.33}} \left(\frac{\mu_w}{\mu_i} \right)^{0.63} \tag{4.11}$$

B_i is the tube-side correction factor and accounts for expansion, contraction and flow reversal of fluid in the exchanger.

Power Loss outside Tubes, E_0

$$-\Delta P_o = B_o \cdot \frac{2 f_o \, G_S^2 \, N_R}{\rho_o} \tag{4.12}$$

N_R is the number of tube rows across which shell fluid flows, f_o is the

167

shell-side friction factor and B_o is the shell-side correction factor for flow reversal, re-crossing of tubes and variations in cross-section.

$$f_o = b_o \left(\frac{\mu_o}{D_o G_S} \right)^{0.15} \tag{4.13}$$

where

$$b_o = 0.23 + \frac{0.11}{(x_T - 1)^{1.08}} \quad \text{for staggered tubes} \tag{4.14a}$$

$$b_o = 0.044 + \frac{0.08 x_T}{(x_T - 1)^{(0.43 + 1.13/x_L)}} \quad \text{for tubes in line} \tag{4.14b}$$

$$E_o = -\frac{\Delta P_o \, G_S \, S_o}{\rho_o \, S_0} = -\frac{\Delta P_o \, G_S \, S_o}{\rho_o \, N_T \, \pi \, D_0 \, L} \tag{4.15}$$

$$S_o = \frac{N_C \, D_C \, L}{n_B} \tag{4.16}$$

Combining (4.12), (4.13), (4.14), (4.15) and (4.16)

$$E_o = \frac{2 B_o \, b_o \, \mu_o^{0.15} \, D_C \, G_S^{2.85} \, N_R \, N_C}{\rho_o^2 \, N_T \, \pi \, D_o^{1.15} \, n_B} \tag{4.17}$$

From the Dittus-Boelter equation

$$\frac{h_o \, D_o}{k_o} = \frac{a_o}{F_S} \cdot \left(\frac{G_S \, D_o}{\mu_o} \right)^{0.8} \left(\frac{Cp_o \, \mu_o}{k_o} \right)^{\frac{1}{3}} \tag{4.18}$$

where $a_o = 0.33$ for staggered tubes and equal to 0.26 for tubes in line. Hence

$$G_S = \left[\frac{h_o \, D_o^{0.4} \, \mu_o^{0.6}}{a_o \, k_o} \cdot \left(\frac{k_o}{Cp_o \, \mu_o} \right)^{\frac{1}{3}} \right]^{1.67} \tag{4.19}$$

From (4.17) and (4.19)

$$E_o = h_o^{4.75} \cdot \frac{2 B_o \, b_o \, \mu_o^{1.42} \, D_o^{0.75} \, D_C \, F_S^{4.75} \, N_R \, N_C}{\rho_o^2 \, N_T \, \pi \, a_o^{4.75} \, n_B \, Cp_o^{1.58} \, k_o^{3.17}} = \Psi_o \, h^{4.75} \tag{4.20}$$

where

$$\Psi_o = \frac{2 B_o \, b_o \, \mu_o^{1.42} \, D_o^{0.75} \, D_C \, F_S^{4.75} \, N_R \, N_C}{\rho_o^2 \, N_T \, \pi \, a_o^{4.75} \, n_B \, Cp_o^{1.58} \, k_o^{3.17}} \tag{4.21}$$

$B_0 = 1.0$ for flow across unbaffled tubes, otherwise, it is equal to the number of tube crosses, generally. Substituting (4.20), (4.10) and (4.3) in (4.1)

$$C_T = S_0 K_F C_{S,0} + \frac{Q.H_Y.C_U}{Cp_U (\Delta T_1 - \Delta T_2 + T_{P_1} - T_{P_2})}$$

$$+ S_0 \Psi_i h_i^{3.5} H_Y C_i + S_0 \Psi_o h_o^{4.75} H_Y C_0 \qquad (4.22)$$

In this equation, S_o, h_i, h_o, and ΔT_2 are the primary variables although only ΔT_2, h_i and h_o, are independent. This is because S_0 is dependent on h_i, h_o, and ΔT_2 as can be seen from the fact that

$$\frac{1}{U_o S_o} = \frac{1}{S_o}\left(\frac{D_o}{D_i h_i} + \frac{1}{h_o} + R_{tw}\right) = \frac{F_T (\Delta T_2 - \Delta T_1)}{Q\ln\left(\dfrac{\Delta T_2}{\Delta T_1}\right)} \qquad (4.23)$$

R_{tw} is the combined resistance of the tube wall and scaling and is given by

$$R_{tw} = \frac{x_w D_o}{k_w D_{lm}} + \frac{D_o}{D_i h_i} + \frac{1}{h_o} \qquad (4.24)$$

h_i and h_o are given by Dittus-Boelter equations (4.8) and (4.18).

Because of the nature of the constraint of (4.23) on (4.22), the optimisation method used by Peters and Timmerhaus was that which employed Lagrange multipliers. Accordingly, the function to be optimised became

$$C_T = S_0 K_F C_{S,0} + \frac{Q.H_Y.C_U}{Cp_U (\Delta T_1 - \Delta T_2 + T_{P_1} - T_{P_2})} + S_0 \Psi_i h_i^{3.5} H_Y C_i$$

$$+ S_0 \Psi_o h_o^{4.75} H_Y C_0 + \lambda \left[\frac{F_T (\Delta T_2 - \Delta T_1)}{Q\ln\left(\dfrac{\Delta T_2}{\Delta T_1}\right)} - \frac{1}{S_o}\left(\frac{D_o}{D_i h_i} + \frac{1}{h_o} + R_{tw}\right)\right] \qquad (4.25)$$

Although the optimum values of heat transfer area, S_o, length of tube, L, and utility flowrate, F_U, are desired, they can be obtained only when the optimum values of U_o, h_i, h_o, ΔT_2, G_s, G_i, S_i, S_o, N_T, N_C, and n_B are determined. Thus, by differentiating (4.25) with respect to any of the variables desired, the results, shown below, were obtained.

4.3.1: Optimum Value of h_o

$$\frac{\partial C_T}{\partial h_o} = 4.75 S_{0,opt} \Psi_o h_{o,opt}^{3.75} H_Y C_0 + \frac{\lambda}{S_{o,opt} h_{o,opt}^2} = 0 \qquad (4.26)$$

$$\frac{\partial C_T}{\partial h_i} = 3.5 S_{0,opt}\ \Psi_i\, h_{i,opt}^{2.5}\, H_Y\, C_i + \frac{\lambda D_o}{S_{o,opt}\, D_i\, h_{i,opt}^2} = 0 \tag{4.27}$$

Eliminating $S_{o,opt}$ and λ from (4.26) and (4.27),

$$h_{o,opt} = \left(\frac{0.74 \Psi_i\, C_i\, D_i}{\Psi_o\, C_o\, D_o}\right)^{0.17} h_{i,opt}^{0.78} \tag{4.28}$$

4.3.2: Optimum Value of h_i

$$\frac{\partial C_T}{\partial S_o} = K_F\, C_{S,0} + \Psi_i\, h_{i,opt}^{3.5}\, H_Y\, C_i + \Psi_o\, h_{o,opt}^{4.75}\, H_Y\, C_0$$

$$+ \frac{\lambda}{S_{o,opt}^2}\left(\frac{D_o}{D_i\, h_{i,opt}} + \frac{1}{h_{o,opt}} + R_{tw}\right) = 0 \tag{4.29}$$

Eliminating $S_{o,opt}$ and λ from (4.29) and (4.27)

$$h_{i,opt}^{3.5}\left[\begin{array}{l} 2.5\,\Psi_i\ H_Y\, C_i + \dfrac{3.5\,\Psi_i\ H_Y\, C_i\, D_i\, R_{tw}\, h_{i,opt}}{D_o} \\[3mm] + 2.9\left(\dfrac{\Psi_i\ C_i\, D_i}{D_o}\right)^{0.83}\left(\Psi_o\, C_0\right)^{0.17}\, H_Y h_{i,opt}^{0.22} \end{array}\right] = K_F\, C_{S,0} \tag{4.30}$$

Equation (4.30) is solved graphically or by a trial and error procedure.

4.3.3: Optimum Value of U_o

This is obtained from (4.23) after having obtained $h_{o,opt}$ and $h_{i,opt}$ from (4.28) and (4.30), respectively. That is

$$U_{o,opt} = \frac{1}{\dfrac{D_o}{D_i\, h_{i,opt}} + \dfrac{1}{h_{o,opt}} + R_{tw}} \tag{4.31}$$

4.3.4: Optimum Value of ΔT_2

$$\frac{\partial C_T}{\partial \Delta T_2} = \frac{Q.H_Y.C_U}{C_{PU}\left(\Delta T_1 - \Delta T_2 + T_{P_1} - T_{P_2}\right)^2} + \frac{\lambda F_T}{Q}\left[\frac{1}{\ln\left(\dfrac{\Delta T_2}{\Delta T_1}\right)} - \frac{\Delta T_1\left(\Delta T_2 - \Delta T_1\right)}{\Delta T_2\left[\ln\left(\dfrac{\Delta T_2}{\Delta T_1}\right)\right]^2}\right] \tag{4.32}$$

170

$$\frac{\partial C_T}{\partial S_o} = K_F \, C_{S,0} + \Psi_i \, h_{i,opt}^{3.5} \, H_Y \, C_i + \Psi_o \, h_{o,opt}^{4.75} \, H_Y \, C_0$$

$$+ \frac{\lambda}{S_{o,opt}^2} \left(\frac{D_o}{D_i \, h_{i,opt}} + \frac{1}{h_{o,opt}} + R_{tw} \right) = 0 \qquad \textit{from (4.29)}$$

Eliminating λ

$$\frac{U_{o,opt} \, F_T \cdot H_Y \cdot C_U}{Cp_U \left(K_F \, C_{S,0} + E_{i,opt} \, H_Y \, C_i + E_{o,opt} \, H_Y \, C_0 \right)}$$

$$= \left(1 + \frac{T_{p,1} - T_{p,2}}{\Delta T_1 - \Delta T_{2,opt}} \right)^2 \cdot \left[\ln \left(\frac{\Delta T_{2,opt}}{\Delta T_1} \right) - 1 - \frac{\Delta T_1}{\Delta T_{2,opt}} \right] \qquad (4.33)$$

4.3.5: Optimum Value of S_o

This is obtained from (4.23) as

$$S_{o,opt} = \frac{Q \ln \left(\dfrac{\Delta T_2}{\Delta T_1} \right)}{F_T \left(\Delta T_2 - \Delta T_1 \right)} \left(\frac{D_o}{D_{i,opt} \, h_{i,opt}} + \frac{1}{h_{o,opt}} + R_{tw} \right) \qquad (4.34)$$

4.3.6: Optimum Value of G_i and G_S

From (4.9)

$$G_{i,opt} = \left[\frac{h_{i,opt} \, D_i^{0.2} \, \mu_i^{0.8}}{0.023 k_i} \cdot \left(\frac{k_i}{Cp_i \, \mu_i} \right)^{\frac{1}{3}} \cdot \left(\frac{\mu_w}{\mu_i} \right)^{0.14} \right]^{1.25} \qquad (4.35)$$

$$G_{S,opt} = \left[\frac{h_{o,opt} \, D_o^{0.4} \, \mu_o^{0.6} \, F_S}{a_o \, k_o} \cdot \left(\frac{k_o}{Cp_o \, \mu_o} \right)^{\frac{1}{3}} \right]^{1.67} \qquad (4.36)$$

4.3.7: Optimum Value of F_U

From (4.3) $\qquad F_{U,opt} = \dfrac{Q}{Cp_U \left(\Delta T_1 - \Delta T_{2,opt} + T_{P_1} - T_{P_2} \right)} \qquad (4.37)$

4.3.8: Optimum Value of S_i and N_T

$$S_{i,opt} = \frac{F_U}{G_{i,opt}} \qquad (4.38)$$

$$N_{T,opt} = \frac{4 n_P S_{i,opt}}{\pi D_i^2} \qquad (4.39)$$

4.3.9: Optimum Value of L

$$L_{opt} = \frac{S_{o,opt}}{\pi D_o N_{T,opt}} \qquad (4.40)$$

4.3.10: Optimum Value of N_C and n_B

For a square pitch and $N_T > 25$

$$N_{C,opt} = 1.37 \left(N_{T,opt} \right)^{0.475} \qquad (4.41)$$

For a triangular pitch and $N_T > 20$

$$N_{C,opt} = 0.94 + \left[\frac{N_{T,opt} - 3.7}{0.907} \right]^{0.5} \qquad (4.42)$$

$$n_{B,opt} = \frac{N_{C,opt} D_C L_{opt}}{S_{o,opt}} \qquad (4.43)$$

4.3.11: A Summary of the Peters and Timmerhaus Procedure

The above results may be utilised in a generalised calculation procedure or algorithm as follows

    1.   Known Values and Conditions

      i   flowrates and necessary temperature changes of the process fluid
      ii   inlet temperature of the utility fluid
      iii   shell and tube exchanger with cross-flow baffling and turbulent flow in both the shell and tube side
      iv   no partial phase changes
      v   all necessary safety factors

2. Values that must be specified or assumed

i   tube diameter, tube wall thickness, tube pitch and arrangement

ii   number of tube passes

iii   heat transfer resistance caused by tube walls, dirt and scale

3. Values assumed at first trial and checked again at optimum conditions

i   average bulk and film temperatures

ii   values of B_i (usually 1.0), B_o/n_B (usually 1.0) and $N_R.Nc/N_T$ (usually 1.0)

4. The Calculation Procedure

i   determine $h_{i,opt}$ from (4.30)

ii   determine h_{0opt} from (4.28)

iii   determine $U_{0,opt}$ from (4.31)

iv   determine $\Delta T_{2,opt}$ from (4.33)

v   determine $S_{0,opt}$ from (4.34)

vi   determine $G_{i,opt}$ and $G_{S,opt}$ from (4.35) and (4.36)

vii   determine F_{Uopt} from (4.37)

viii   determine $S_{i,opt}$ and $N_{T,opt}$ from (4.38) and (4.39)

ix   determine L_{opt} from (4.40)

x   determine $N_{C,opt}$ and $n_{B,opt}$ from (4.41) or (4.42) and (4.43)

xi   check the assumptions and, if any are invalid, make new ones and repeat the procedure

4.3.12: Simplifications of the Method of Peters and Timmerhaus (1985)

Peters and Timmerhaus (1985) listed cases where simplifications to this general procedure can be made. These are:

A: Power cost on the shell or tube side immaterial

1. Shell-side power cost immaterial ($C_o = 0$ and h_o = constant)

Then, $h_{i,opt}$ is given by

$$h_{i,opt}^{3.5}\left[2.5\,\Psi_i\,H_Y\,C_i + \frac{3.5\,\Psi_i\,H_Y\,C_i\,D_i}{D_o}\left(\frac{1}{h_o}+R_{tw}\right)h_{i,opt}\right]=K_F\,C_{S,0} \quad (4.44)$$

2. <u>Tube-side power cost immaterial ($C_i = 0$ and h_i = constant)</u>

$h_{o,opt}$ is obtained from

$$h_{o,opt}^{4.75}\left[3.75\,\Psi_o\,H_Y\,C_o + 4.75\,\Psi_o\,H_Y\,C_o\left(\frac{D_o}{D_i\,h_i}+R_{tw}\right)h_{o,opt}\right]=K_F\,C_{S,0} \quad (4.45)$$

B: <u>Power costs on both the shell- and tube-side important</u>

Then $\quad\dfrac{Power\,cost\,on\,tube\,side}{Power\,cost\,on\,shell\,side}=\dfrac{E_i\,C_i}{E_o\,C_o}=\dfrac{\Psi_i\,h_i^{3.5}\,C_i}{\Psi_o\,h_o^{4.75}\,C_o}\quad (4.46)$

C: <u>Velocity of one fluid fixed (h_i or h_o fixed)</u>

1. <u>Velocity inside tube is fixed or h_i is constant</u>

$h_{o,opt}$ is obtained by trial and error from

$$h_{o,opt}^{4.75}\left[3.75\,\Psi_o\,H_Y\,C_o + 4.75\,\Psi_o\,H_Y\,C_o\left(\frac{D_o}{D_i\,h_i}+R_{tw}\right)h_{o,opt}\right]$$
$$= K_F\,C_{S,0} + \Psi_i\,h_i^{3.5}\,H_Y\,C_i \quad (4.47)$$

2. <u>Velocity outside tube is fixed or h_o is constant</u>

$h_{i,opt}$ is obtained by trial and error from

$$h_{i,opt}^{3.5}\left[2.5\,\Psi_i\,H_Y\,C_i + \frac{3.5\,\Psi_i\,H_Y\,C_i\,D_i}{D_o}\left(\frac{1}{h_o}+R_{tw}\right)h_{i,opt}\right]$$
$$= K_F\,C_{S,0} + \Psi_o\,h_o^{4.75}\,H_Y\,C_o \quad (4.48)$$

4.4: Optimisation of Heat Exchanger Design by the Method of Jegede and Polley (1992)

The basic procedure, in this method, is to develop, for each stream, a general expression which relates the pressure drop of the stream to both the area of the exchanger and the heat transfer ooefficient of the stream. This expression can, then, be applied to any heat exchanger provided the applicable friction factor and heat transfer equations are known.

4.4.1: General Pressure Drop Relationship

For any heat exchanger system, the pressure drop, ΔP, and the fluid velocity, u, are related as follows

$$\Delta P = K_{P,1} \, L u^{2-n} \tag{4.49}$$

where n is the Reynold's number exponent in the friction factor equation, L is a characteristic length and $K_{P,1}$ is a constant. The length, L, can be related to the heat transfer area, S, the volumetric throughput, V_0, and fluid velocity, u, as

$$L = K_{P,2} \, S \frac{u}{V_0} \tag{4.50}$$

where $K_{P,2}$ is another constant. The heat transfer coefficient, h, can be related to the fluid velocity, u, as

$$h = K_{P,3} \, u^m \tag{4.51}$$

where m is the Reynold's number exponent in the heat transfer equation and $K_{P,3}$ is a constant.

Combining (4.49), (4.50) and (4.51)

$$\Delta P = \frac{K_{P,4}}{V_0} \cdot S h^{\left(\frac{3-n}{m}\right)} \tag{4.52}$$

where $K_{P,4}$ is another constant.

This, equation (4.52), is the general expression required.

4.4.2: Application to a Shell and Tube Heat Exchanger

1. Tube-Side Relationships

   a. Pressure drop - Velocity Relationship

$$\Delta P_i = 2 f_i \cdot \frac{L}{D_i} \cdot n_P \rho_i \, u_i^2 \tag{4.53}$$

where n_P is the number of tube passes. When $f = 0.046 \, Re^{-0.2}$

$$\Delta P_i = K_{P,1}\, L n_P\, u_i^{1.8} \qquad (4.54)$$

where
$$K_{P,1} = 0.092\left(\frac{\rho}{D_i}\right)\cdot\left(\frac{\mu}{\rho D_i}\right)^{0.2}. \qquad (4.55)$$

b. Length - Velocity Relationship

Heat transfer area, $S_S = N\pi D_o L$ $\qquad\qquad (4.56)$

Volumetric flowrate,
$$V_o = \frac{N\pi D_i^2\, u_i}{4 n_T} \qquad (4.57)$$

From (4.56) and (4.57),
$$L = \frac{S_S\, D_i^2\, u_i}{4 n_T\, D_o\, V_o} \qquad (4.58)$$

From (4.58) and (4.54),
$$\Delta P_i = K_{P,1}\frac{S_S\, D_i^2\, u_i^{2.8}}{4 D_o\, V_o} \qquad (4.59)$$

c. Heat Transfer Coefficient - Velocity Relationship

From the Dittus-Boelter equation

$$h_i = 0.023\left(\frac{k}{D_i}\right).\Pr^{\frac{1}{3}}.\mathrm{Re}^{0.8} \qquad (4.60)$$

This can be re-arranged to give $\qquad h_i = K_2\, u_i^{0.8} \qquad (4.61)$

where
$$K_2 = 0.023\left(\frac{k}{D_i}\right).\Pr^{\frac{1}{3}}.\left(\frac{\rho D_i}{\mu}\right)^{0.2} \qquad (4.62)$$

k is the thermal conductivity of the fluid.

Substituting for u, in (4.61) and (4.59)
$$\Delta P_i = K_{P,T}.S_S.h_i^{3.5} \qquad (4.63)$$

where
$$K_{P,T} = K_{P,1}\frac{D_i^2}{4 D_o\, V_o\, K_2^{3.5}} \qquad (4.64)$$

2. Shell - Side Relationships

a. Pressure Drop - Velocity Relationship

From Kern's correlations for a 25% baffle cut

$$\Delta P_S = f_S \cdot \frac{D_S \left(N_B + 1\right) G_S^2}{2 \rho D_E} \qquad (4.65)$$

where D_S is the shell diameter, N_B is the number of baffles, G_S is the mass flow velocity per unit cross-sectional area, D_E is the equivalent diameter for flow through the shell and

$$f_S = 1.79 . \mathrm{Re}_S^{-0.19} \qquad (4.66)$$

$$\mathrm{Re}_S = \frac{D_E G_S}{\mu_S} \qquad (4.67)$$

Substituting (4.67) and (4.66) into (4.65)

$$\Delta P_S = K_{S,1} . D_S \left(N_B + 1\right) u_S^{1.81} \qquad (4.68)$$

where $\qquad K_{S,1} = 1.79 . \left(\frac{\mu_S}{D_E}\right)^{3.81} . \frac{\rho^{1.8}}{2 \rho D_E} \qquad (4.69)$

and u_S is the shell-side velocity.

b. Shell Diameter - Velocity Relationship

Heat transfer Area: $\quad S_S = N\pi D_o L = N\pi D_o \left(N_B + 1\right) B_S \qquad (4.70)$

Tube count: $\qquad\qquad N = \frac{\pi D_S^2}{4 P_T^2} \qquad (4.71)$

B_S is the baffle spacing and P_T is the tube pitch.

Combining (4.71) and (4.70)

$$\frac{S_S}{\pi D_o \left(N_B + 1\right) B_S} = N = \frac{\pi D_S^2}{4 P_T^2} \qquad (4.72)$$

The shell fluid flow area is given by

$$a_S = \frac{D_S}{P_T} \left(P_T - D_o\right) B_S \qquad (4.73)$$

If the shell fluid volumetric flowrate is V_S, then the shell fluid velocity is given by

$$u_S = \frac{V_S}{a_S} = \frac{V_S}{\dfrac{D_S}{P_T} \left(P_T - D_o\right) B_S} \qquad (4.74)$$

From which $\qquad D_S B_S = \frac{V_S P_T}{u_S \left(P_T - D_o\right)} \qquad (4.75)$

From (4.72)
$$D_S \, B_S = \frac{4 P_T^2 \, S_S}{\pi^2 \, D_o \left(N_B + 1\right) D_S}$$
(4.76)

We get that
$$\frac{V_S \, P_T}{u_S \left(P_T - D_o\right)} = \frac{4 P_T^2 \, S_S}{\pi^2 \, D_o \left(N_B + 1\right) D_S}$$
(4.76)

That is
$$\left(N_B + 1\right) D_S = \frac{4 P_T \, S_S \, u_S \left(P_T - D_o\right)}{\pi^2 \, D_o \, V_S} = K_{S,2} \, S_S \, u_S$$
(4.77)

where
$$K_{S,2} = \frac{4 P_T \left(P_T - D_o\right)}{\pi^2 \, D_o \, V_S}$$
(4.78)

c. Heat Transfer Coefficient – Velocity Relationship

From Kern's equation
$$j_h = 0.36 \, \mathrm{Re}_S^{0.55}$$
(4.79)

This gives
$$h_S = K_{S,3} \, u_S^{0.55}$$
(4.80)

where
$$K_{S,3} = 0.36 \cdot \left(\frac{k}{D_E}\right) \mathrm{Pr}^{\frac{1}{3}} \cdot \left(\frac{\rho D_E}{\mu_S}\right)^{0.55}$$
(4.81)

Combining (4.69), (4.78) and (4.81), we get
$$\Delta P_S = K_{P,S} \cdot S_S \, h_S^{5.1}$$
(4.82)

where
$$K_{P,S} = K_{P,1} \cdot K_{P,2} \cdot K_{P,3}$$
(4.83)

3: Recommended Design Algorithm when Pressure Drops are Specified

Optimisation of heat exchanger designs, in which the allowable pressure drop is specified before hand, is handled by the methods of Peters and Timmerhaus (1985) and others, of which the method of Peters and Timmerhaus (1985) is the most comprehensive. Jegede and Polley (1992), however, point out, correctly, that the optimum obtained in this way cannot be a true optimum because the set allowable pressure drop is not, necessarily, optimum, unless in a constraint situation.

Their method, which can determine the optimum pressure drop, is claimed to be more straightforward than other methods even for situations in which the allowable pressure drop is specified *a - priori*. In their method, illustrated below, the whole exchanger is defined by the

178

three equations

$$\Delta P_i = K_{P,T} \cdot S_S \cdot h_i^{3.5} \qquad \qquad from \quad (4.63)$$

$$\Delta P_S = K_{P,S} \cdot S_S \, h_S^{5.1} \qquad \qquad from \quad (4.82)$$

$$Q = U \cdot S \cdot \Delta T_{lm} \qquad \qquad (4.84)$$

where

$$\frac{1}{U} = \frac{1}{h_i} + \frac{1}{h_S} + R_f \qquad \qquad (4.85)$$

Since ΔP_i, ΔP_S, Q and ΔT_{lm} are design requirements, there are three equations, with three unknowns, h_i, h_S, and S, which can, easily, be solved for, as shown below.

Thus, from (4.63)

$$h_i^{3.5} = \frac{\Delta P_i}{K_{P,T} \cdot S_S} . \qquad \qquad (4.86)$$

From (4.85) and (4.84)

$$S_S = \frac{Q}{\Delta T_{lm}} \left(\frac{1}{h_i} + \frac{1}{h_S} + R_f \right) \qquad \qquad (4.87)$$

From (4.86) and (4.87)

$$h_i = \left[\frac{\Delta P_i \, \Delta T_{lm}}{K_{P,T} \cdot Q \cdot \left(\dfrac{1}{h_i} + \dfrac{1}{h_S} + R_f \right)} \right]^{\frac{1}{3.5}} . \qquad \qquad (4.88)$$

From (4.63) and (4.82)

$$\frac{1}{h_S} = \left[\frac{\Delta P_i}{\Delta P_S} \cdot \frac{K_{P,S}}{K_{P,T} \cdot h_i^{3.5}} \right]^{\frac{1}{5.1}} . \qquad \qquad (4.89)$$

Equation (4.89) can be substituted into (4.88) which is, then, solved, by convergence methods, to determine the optimum h_i, The optimum h_i, is, then, used to obtain the optimum h_S and, hence, the optimum A_S. With these, other optimal exchanger variables are obtained as

Tube-side

From (4.61)

$$u_i = \left(\frac{h_i}{K_2} \right)^{\frac{1}{0.8}} \qquad \qquad (4.90)$$

From (4.57)

$$N = \frac{4 \, n_T \, V_o}{\pi \, D_i^2 \, u_i} \qquad \qquad (4.91)$$

179

From (4.56)
$$L = \frac{S_S}{N \pi D_o}$$
(4.92)

Shell -side

From (4.80)
$$u_S = \left(\frac{h_S}{K_{S,3}} \right)^{\frac{1}{0.55}}$$
(4.93)

From (4.72)
$$D_S = \left(\frac{4 P_T^2}{\pi N} \right)^{\frac{1}{2}}$$
(4.94)

From (4.75)
$$B_S = \frac{V_S P_T}{u_S D_S (P_T - D_o)}$$
(4.95)

Also
$$N_B = \frac{1}{B_S} - 1$$
(4.96)

4: Recommended Design Algorithm when Allowable Pressure Drops are Unknown

Jegede and Polley (1992) considered this situation to be of more general occurrence and recommended that objective selection of allowable pressure drops be made. The choice will depend on the relative cost of power and capital.

Their recommended optimisation procedure recognised two degrees of freedom defined by (i) h_i (or h_S) and S_S or (ii) h_i and h_S, which, if any two are known, determine the third and hence the system.

For example, if h_i and S_S are chosen, the optimum heat exchanger is obtained when

$$\frac{\partial C_T}{\partial S_S} = 0 \qquad \text{and} \qquad \frac{\partial C_T}{\partial h_i} = 0$$
(4.97)

If h_i and h_S are chosen, the optimum heat exchanger is obtained when

$$\frac{\partial C_T}{\partial h_S} = 0 \qquad \text{and} \qquad \frac{\partial C_T}{\partial h_i} = 0$$
(4.98)

C_T is the total capital and operating cost of the exchanger, estimated as in the method of Peters and Timmerhaus (1985) but expressed solely in terms of h_i, h_S and S_S.

Jegede and Polley (1992) give the capital costs, in U.S.A. dollars, of heat exchangers, pumps and compressors as follows:

Heat Exchangers $\quad C_{exchanger} = 19,600 + 1008S_S^{0.81}$ $\qquad$ (4.99)

Pumps $\qquad C_{pump} = 1,410 + 4.64(V.\Delta P)^{0.68}$ $\qquad$ (4.100)

Compressors $\qquad C_{compr} = 5,434(HP)^{0.774}$ $\qquad$ (4.101)

4.5: Heat Exchanger Networks

Energy considerations dominate the analysis and operation of industrial and process plants and have led to the use of heat exchanger networks, where more than one or few heat exchangers are required to implement the heat exchange duty in the plant.

Heat exchanger networks (HENs) are systems of heat exchangers arranged in such a way that heat energy is utilised most economically within a process. Fluids, which enter at very high or very low temperatures, are passed, successively, through different exchangers such that very little heat energy is carried away unused. The aim is to find a configuration of heat exchangers which accomplishes the required heat exchange at the lowest total cost. The three approaches, which have been employed, are summarised below.

i.   Determine the Minimum Heat Exchanger (Rudd *et al.* 1973)

Since the cost of a heat exchanger varies inversely as the temperature difference, a point is reached, in the use of a single exchanger, where it is more economical to add another exchanger than to continue to increase the size (surface area) of the single exchanger. This point is seen, from experience, to be at the minimum temperature difference between the streams exchanging heat energy (approach temperature) of between 8 C and 10 C. Thus, heat exchange between two streams is arranged to ensure that the temperature difference at either end is not less than the minimum approach temperature.

ii. Stay as close as possible to Ambient Temperatures (Rudd *et al.* 1973)

It is known, from experience, that processing costs increase with excursions from ambient temperature. Also, heating and cooling fluids, such as steam and water, tend to be available at characteristic temperatures. Schemes which allow heating and cooling close to these characteristic temperatures are, therefore, advisable.

Hence, subject to the minimum approach temperature, heat exchange should be between either, the hottest stream to be cooled and the warmest stream to be heated, or between the coldest stream to be heated with the coolest stream to be cooled.

iii.   Determine the Relative Importance of Thermal versus Mechanical Energy

An important factor in heat exchanger optimisation is the ratio of thermal to mechanical energy. For high density fluids, such as water, the friction power expenditure (the mechanical energy spent in pumping fluid through the heat exchanger) is small relative to the heat transfer rate and, therefore, is not the controlling factor. For low density fluids, such as air, however, friction power expenditure is, often, as much as the heat transferred (Kays & London, 1964).

Note, however, that, in general, the heat transfer rate is directly proportional to the fluid flow velocity while the friction power expenditure is proportional to the square (or cube) of the fluid flow velocity. In low density fluids, therefore, the friction power expenditure can be reduced by reducing the fluid flow velocities without, appreciably, reducing the heat transfer rate. Although optimisation of HENs is quite complex and more heuristic than analytical, a pattern or history of optimisation of HENs has emerged. A version of this pattern, due to Kleinschodt and Hammer (1983), is given in Figs 6.1 to 6.4 below.

Figure 4.1A shows the first generation HEN which used close cold end approach temperatures. The major drawback of this approach was that these HENs required high shell counts and large surface areas.

Figure 4.1B shows that the second generation cyclic network solved the problem of close approaches and high shell count but created the problem of higher piping costs as a result of the intermeshing of services.

Figure 4.2A shows the third generation HEN in which all exchangers were put in parallel. This arrangement is best when all hot streams are at the same temperature. It is, however, essentially, a first generation HEN and, therefore, not an improvement over it.

Figure 4.2B shows the current optimal method of HEN integration. It minimises shell count, surface area and service count and is based on, alternately, splitting and cycling the heat exchangers.

Figure 4.1: First and Second Generation Heat Exchanger Networks (Kleinschodt & Hammer. 1983)

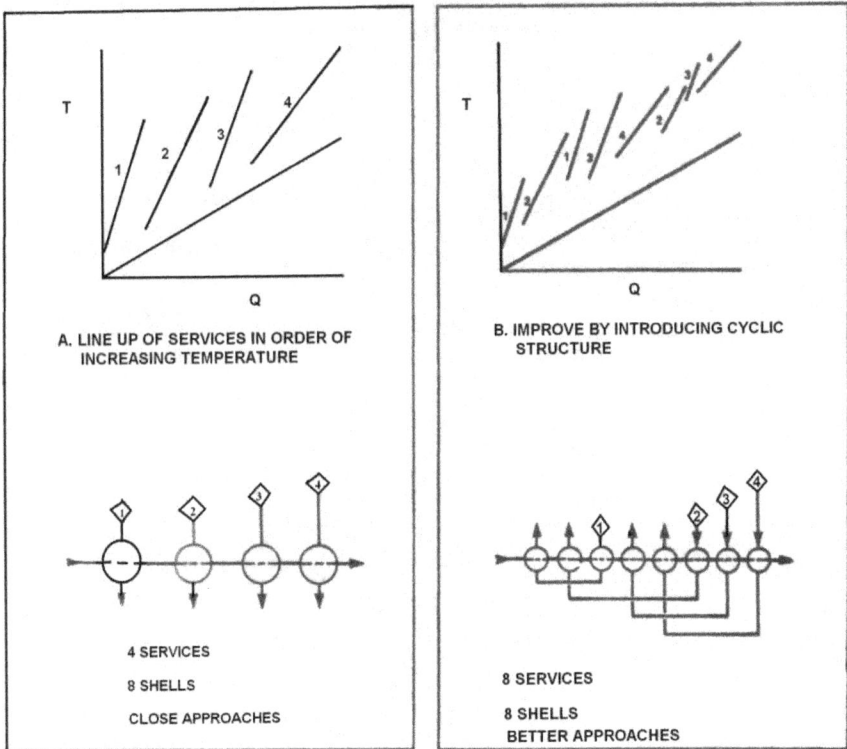

Figure 4.2: Third and Current Generation Heat Exchanger Networks (Kleinschodt & Hammer. 1983)

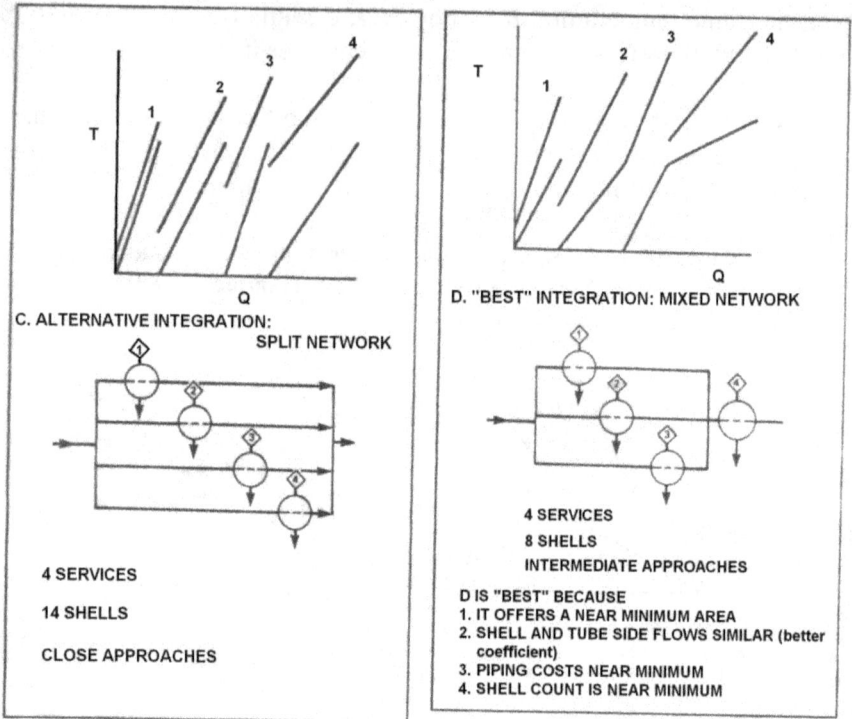

C. ALTERNATIVE INTEGRATION: SPLIT NETWORK

4 SERVICES

14 SHELLS

CLOSE APPROACHES

D. "BEST" INTEGRATION: MIXED NETWORK

4 SERVICES

8 SHELLS

INTERMEDIATE APPROACHES

D IS "BEST" BECAUSE
1. IT OFFERS A NEAR MINIMUM AREA
2. SHELL AND TUBE SIDE FLOWS SIMILAR (better coefficient)
3. PIPING COSTS NEAR MINIMUM
4. SHELL COUNT IS NEAR MINIMUM

References for Chapter Four

1   Jegede F O. and Polley G. T. (1992); *Optimum Heat Exchanger Design*; Trans. Inst. Chem. Engrs.; Part A - Chem. Eng. Res. & Dev., Vol. 70 No. A2; March; Brit. Inst. Chem. Engrs.; Rugby, U.K.

2   Jenssen S. K. (1969); *Heat Exchanger Optimisation*; Chem. Engrs. Prog; in Jegede and Polley (1992); *Optimum Heat .Exchanger Design*; Trans. Inst. Chem. Engrs.; Part A - Chem. Eng. Res. & Dev., Vol. 70 No. A2; March; Brit. Inst. Chem. Engrs.; Rugby, U.K.

3   Kays W. M. and London A. L.; *Compact Heat Exchangers*, 2nd. Edn.; McGraw-Hill Book Co.; New York, 1954

4   Kleinschodt F. J. and Hammer G. A. (1983); Chem. Engrs. Prog.; Vol 79, July; AIChE; N. Y., U.S.A

5   McAdams W. H.; *Heat Transmission*; in Jegede and Polley (1992); *Optimum Heat .Exchanger Design*; Trans. Inst. Chem. Engrs.; Part A - Chem. Eng. Res. & Dev., Vol. 70 No. A2; March; Brit. Inst. Chem. Engrs.; Rugby, U.K.

6   Pase G. K. (.1986); *Computer Programs for Heat Exchanger Design*; Chem. Engrs. Prog.; Vol. 89 No. 2, September; AIChE; N. Y., U.S.A.

7   Peters M. S. and Timmerhaus K. D.; *Plant Design and Economics for Chemical Engineers*; 3rd Edn, International Student Edn.; McGraw-Hill International Book Co.; London, Tokyo, 1985.

8   Sargent R. W. H.; Class Notes in Heat Transfer; Imp. Coll. London, U. K.; 1965

9   Steinmeyer D. E. (1976); *Energy Price Impacts Design*; Hydrocarbon Process; in Jegede and Polley (1992); *Optimum Heat .Exchanger Design*; Trans. Inst. Chem. Engrs.; Part A - Chem. Eng. Res. & Dev., Vol. 70 No. A2; March; Brit. Inst. Chem. Engrs.; Rugby, U.K.

10   Steinmeycr D. E.(1982); *Take your Pick: Capital or Energy*; CHEMTECH; March; 188; A.C.S.; U.S.A.

11   www.lauterbach.com

APPENDIX I
Derivation of the Navier-Stokes Equations

The derivation of the Navier-Stokes equation is, often, bewildering to the beginner. This is partly because the derivation has to conform to the requirements of fluid mechanics and mathematics to use exact terminology or else the rigour is lost. The difficulty encountered by beginners will depend on their level of familiarity with scalars, vectors or tensors.

Whether you are familiar with vectors and tensors or not, the bottom line is that the Navier-Stokes equation is based on the application of Newton's second law of motion to fluids. This law states that

$$Force = massx\, acceleration \qquad (I.1)$$

We can see that, because fluids can be deformed, the estimation of force, mass and acceleration will be a bit more complicated than for a rigid solid. To simplify this process, the first step is to make a small number of basic assumptions. These are that

1. The fluid, at the point of interest, is a continuum. In other words, it is not composed of discrete particles in this region of interest.
2. All fluid properties such as pressure, velocity, density, temperature, viscosity, etc, are differentiable at any point in the fluid, even if weakly.

Because the fluid may or may not be in motion, it is necessary to state clearly whether the analysis of fluid motion is based on an observer moving with the fluid (the Lagrangian perspective) or observing the fluid motion from a fixed point in space (the Eulerian perspective). This affects the mathematics and the subsequent result.

Because the fluid is associated with mass, momentum and energy, the derivation can be seen as one in which these are conserved. You may wish to recall that an analysis, in which some variable or property is conserved, is one in which we accept that, for each of such a variable or property,

$$Input + Generation\ Consumption = Output + Accumulation \qquad (I.2)$$

You may, also, recall that equation (I.2) is useful if applied to a finite, even if arbitrary volume, called the control volume.

Our experience with fluids indicates that the forces involved in fluid motion are the so called body forces (gravity), pressure gradient forces, viscous forces and inertia forces.

187

Finally, mathematics enables us to define fluid velocity, fluid acceleration and other variables, in terms of the time and space coordinates of the system. Many of you are familiar with the common coordinate systems: the rectangular or Cartesian coordinate system, the polar or cylindrical coordinate system and the spherical coordinate system.

Derivation of the Navier-Stokes Equations in Vector Notation

In this derivation, based on that from Wikipedia (2009), equation (I.2) is stated in a more explicit and useful form as the Reynold's transport theorem. This theorem states that the sum of changes of some intensive property, in a control volume, must be equal to what is gained or lost through the boundaries of the control volume plus what is created or consumed by sources and sinks inside the control volume. If this intensive property is represented by Φ and the control volume by Ω, the Reynold's transport theorem states that

$$\frac{d}{dt}\int_{\Omega} \Phi dV = -\int_{\partial\Omega} \Phi v.nd\,A - \int_{\Omega} Q dV \qquad (I.3)$$

V is the volume variable, v is fluid velocity and Q represents the sources and sinks within the control volume.

From mathematics, the divergence theorem can be applied to a surface integral (the second term in equation (I.3)) in order to change it to a volume integral. Thus equation (I.3) becomes

$$\frac{d}{dt}\int_{\Omega} \Phi dV = -\int_{\Omega} \nabla.(\Phi v)dV - \int_{\Omega} Q dV \qquad (I.4)$$

From Leibniz's rule

$$\frac{\partial}{\partial t}\int_{\Omega} \Phi dV = \int_{\Omega} \frac{\partial \Phi}{\partial t} dV \qquad (I.5)$$

Combining equations (I.4) and (I.5)

$$\int_{\Omega}\left(\frac{\partial \Phi}{\partial t} + \nabla.(\Phi v) + Q\right)dV = 0 \qquad (I.6)$$

Equation (I.6) is true only if

$$\frac{\partial \Phi}{\partial t} + \nabla.(\Phi v) + Q = 0 \qquad (I.7)$$

Conservation of Momentum

If the intensive property of the fluid under consideration is momentum, usually, expressed as ρv, equation (I.7) may be rewritten as

$$\frac{\partial \rho v}{\partial t} + \nabla.(\rho v v) + Q = 0 \qquad (I.8)$$

Remembering that v is a vector, equation (I.8) becomes,

$$\frac{\partial \rho}{\partial t} v + \rho \frac{\partial v}{\partial t} + \nabla(\rho v).v + \rho v \nabla. v = b \qquad (I.9)$$

where b, now, represents sources or sinks of momentum per unit volume. Equation (I.9) can be, further, expressed as

$$\frac{\partial \rho}{\partial t} v + \rho \frac{\partial v}{\partial t} + \nabla(\rho)v.v + \rho \nabla(v).v + v \rho \nabla.v = b \qquad (I.10)$$

which can be even further expressed as

$$v \frac{\partial \rho}{\partial t} + \rho \frac{\partial v}{\partial t} + vv.\nabla \rho + \rho v.\nabla v + \rho v \nabla.v = v \qquad (I.11)$$

Equation (I.11) can be rearranged to

$$v \left(\frac{\partial \rho}{\partial t} + v.\nabla \rho + \rho \nabla.v \right) + \rho \left(\frac{\partial v}{\partial t} + v.\nabla v \right) = b \qquad (I.12)$$

From vector algebra

$$v.\nabla \rho + \rho \nabla.v = \nabla.(\rho v) \qquad (I.13)$$

Hence, from (I.12) and (I.13),

$$v \left(\frac{\partial \rho}{\partial t} + \nabla.(\rho v) \right) + \rho \left(\frac{\partial v}{\partial t} + v.\nabla v \right) = b \qquad (I.14)$$

Since mass is conserved

$$v \left(\frac{\partial \rho}{\partial t} + \nabla.(\rho v) \right) = 0 \qquad (I.15)$$

and

$$\rho \left(\frac{\partial v}{\partial t} + v.\nabla v \right) = \rho \frac{Dv}{Dt} = b \qquad (I.16)$$

Conservation of Mass

If the intensive property of the fluid under consideration is mass, represented by density, ρ, there are no sources or sinks so that Q is zero, equation (I.7) may be rewritten as

$$\frac{\partial \rho}{\partial t} + \nabla . (\rho v) = 0 \qquad (I.17)$$

Equation (I.17) is the mass continuity equation, generally, referred to as the continuity equation, for short. For incompressible fluids, ρ is constant and

$$\nabla . \bar{v} = 0 \qquad (I.18)$$

Volume is conserved.

Conservation of Sources and Sinks, **b**, as Forces

An analysis of the forces acting on an infinitesimal cube of fluid shows these forces to be made up of stresses, (direct, σ, and tangential, τ) and of gravity or body forces, BF such that

$$b = \nabla . \sigma + BF \qquad (I.19)$$

We can, then, state equation (I.16), in its most general form, as

$$\rho \left(\frac{\partial v}{\partial t} + v . \nabla v \right) = \rho \frac{Dv}{Dt} = \nabla . \sigma + BF \qquad (I.20)$$

where

$$\sigma_{ij} = \begin{pmatrix} \sigma_{xx} & \tau_{xy} & \tau_{xz} \\ \tau_{yx} & \sigma_{yy} & \tau_{yz} \\ \tau_{zx} & \tau_{zy} & \sigma_{zz} \end{pmatrix} = - \begin{pmatrix} p & 0 & 0 \\ 0 & p & 0 \\ 0 & 0 & p \end{pmatrix} + \begin{pmatrix} \sigma_{xx} + p & \tau_{xy} & \tau_{xz} \\ \tau_{yx} & \sigma_{yy} + p & \tau_{yz} \\ \tau_{zx} & \tau_{zy} & \sigma_{zz} + p \end{pmatrix}$$

$$= -pI + T \qquad (I.21)$$

I is a 3 x 3 identity matrix, T is the stress tensor and p is the pressure minus the normal stress and is given by

$$p = \frac{1}{3} \left(\sigma_{xx} + \sigma_{yy} + \sigma_{zz} \right) \qquad (I.22)$$

The Navier-Stokes equation may, now, be stated in its general form, as

$$\rho \frac{Dv}{Dt} = -\nabla p + \nabla . T + BF \qquad (I.23)$$

The Navier-Stokes Equation for a Newtonian Fluid

For a Newtonian fluid, the stress tensor is a linear function of the strain rates. The fluid is isotropic and when the fluid is at rest, $\nabla . T = 0$ so that pressure is hydrostatic. For such a fluid

$$T_{ij} = \mu \left(\frac{\partial u_i}{\partial x_j} + \frac{\partial u_j}{\partial x_i} \right) + \delta_{ij} \lambda \nabla.v = 0 \qquad (I.24)$$

δ_{ij} is the Kronecker delta function, μ is the Newtonian, or first coefficient of, viscosity and λ second coefficient of viscosity. λ is negligible in some situations and not so in other situations when it takes on the value $\lambda = - 2\mu/3$. The gravitational body force, $BF = \rho g$.

Substituting equation (I.24) into (I.23) gives the Navier-Stokes equation for a Newtonian fluid as

$$\rho \frac{Dv}{Dt} = -\nabla p + \nabla. \left(\mu. \left(\nabla v + (\nabla v)^T \right) \right) + \nabla (\lambda \nabla.v) + \rho g \qquad (I.25)$$

with equation (I.17) still the associated mass continuity equation.

For incompressible fluids, μ is constant and $\lambda = 0$ so that

$$\rho \frac{Dv}{Dt} = -\nabla p + \mu \nabla^2 v + \rho g \qquad (I.26)$$

Conservation of Energy

For the conservation of heat energy, represented by the enthalpy, h, an equation of state will be needed. If the ideal gas law is appropriate, the conservation of energy will give

$$\rho \frac{Dh}{Dt} = \frac{Dp}{Dt} + \nabla. (k \nabla T) + \Psi \qquad (I.27)$$

where T is the temperature and Ψ is energy dissipated as a result of viscous effects.

Derivation of the Navier-Stokes Equations by means of a Physical Model

This derivation, partly based on that by Juster (GLY 5932), considers a small fluid element with dimensions, in Cartesian co-ordinates, of Δx, Δy and Δz, such as that shown below:

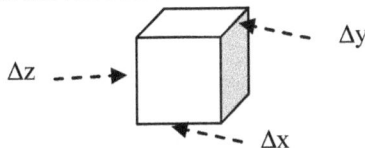

For such an element of fluid to be in motion, forces must act on it, according to Newton's second law of motion which states, as we saw earlier, that

$$Force = mass \times acceleration \qquad (I.1)$$

The obvious forces we would expect to act on it would be those due to gravity (body forces), mechanical pressure (pressure gradients) and viscosity (viscous stresses). An equation of motion of a fluid may, therefore, be obtained if Newton's second law of motion can be evaluated for it. Since the fluid is in a control volume, a force balance according to equation (I.2) enables us to evaluate the resultant forces to be used in the Newton's second law equation.

The most obvious place to look for these forces would be the surfaces of the fluid element. These forces would be stresses and would consist of direct and tangential stresses, as in elastic solids. Some of them are shown, to minimise confusing the beginner, in the diagram below.

Direct and Tangential stresses

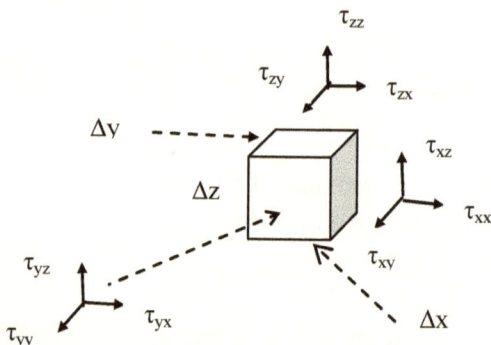

For direct stresses, the two subscripts are the same and represent the direction in which the stress acts. For tangential stresses, the first subscript represents the direction perpendicular to the direction of the stress while the second subscript represents the direction of the stress. Thus τ_{yx} represents the tangential stress acting in the x-direction on the xz plane while τ_{zy} is the stress acting in the y-direction on the xy plane. Simple permutation will show that there would be six such tangential stresses and three direct stresses. These comprise the viscous stress tensor:

$$\tau_{ij} = \begin{pmatrix} \tau_{xx} & \tau_{xy} & \tau_{xz} \\ \tau_{yx} & \tau_{yy} & \tau_{yz} \\ \tau_{zx} & \tau_{zy} & \tau_{zz} \end{pmatrix} \qquad (I.28)$$

If we take the centre of any face of our elemental parallelpiped to be the point of action of any force we wish to consider, we can see, for example, on the $\Delta z\Delta y$ and $\Delta x\Delta y$ planes, that the pressure and direct viscous forces would be

For the $\Delta z\Delta y$ plane,

$$P_{x1} = P_{x0} - \frac{\Delta x}{2}\frac{\partial p_x}{\partial x}, \quad P_{x2} = P_{x0} + \frac{\Delta x}{2}\frac{\partial p_x}{\partial x}, \quad \tau_{xx1} = \tau_{xx0} - \frac{\Delta x}{2}\frac{\partial \tau_{xx}}{\partial x} \quad \text{and}$$

$$\tau_{xx2} = \tau_{xx0} + \frac{\Delta x}{2}\frac{\partial \tau_{xx}}{\partial x},$$

For the $\Delta x\Delta y$ plane,

$$P_{z1} = P_{z0} - \frac{\Delta z}{2}\frac{\partial p_z}{\partial z}, \quad \tau_{zz1} = \tau_{zz0} - \frac{\Delta z}{2}\frac{\partial \tau_{zz}}{\partial z} \quad \text{at one end}$$

and $P_{z2} = P_{z0} + \frac{\Delta z}{2}\frac{\partial p_z}{\partial z}, \quad \tau_{zz2} = \tau_{zz0} + \frac{\Delta z}{2}\frac{\partial \tau_{zz}}{\partial z}$ at the other end.

We have defined the pressures and the direct stresses, P_x, P_z, τ_{xx} and τ_{zz} at the center as P_{x0}, P_{z0}, τ_{xx0} and τ_{zz0} and used only linear variations to estimate them at any distance we choose, in this case $(\Delta z)/2$ and $(\Delta x)/2$.

The net force in the x-direction, due to pressure, is then

$$F_{net\,pressure} = P_{x1}.\Delta z.\Delta y - P_{x2}.\Delta z.\Delta y$$

$$= \left(P_{x0} - \frac{\Delta x}{2}\frac{\partial P_x}{\partial x} - P_{x0} - \frac{\Delta x}{2}\frac{\partial P_x}{\partial x} \right).\Delta z.\Delta y = -\frac{\partial P_x}{\partial x}\Delta x.\Delta z.\Delta y \quad (I.29)$$

The net force in the x-direction, due to the direct stress, is then

$$F_{net\,direct\,stress} = \tau_{xx2}.\Delta z.\Delta y - \tau_{xx1}.\Delta z.\Delta y$$

$$= \left(\tau_{xx0} + \frac{\Delta x}{2}\frac{\partial \tau_{xx}}{\partial x} - \tau_{xx0} + \frac{\Delta x}{2}\frac{\partial \tau_{xx}}{\partial x} \right).\Delta z.\Delta y$$

$$= \frac{\partial \tau_{xx}}{\partial x}\Delta x.\Delta z.\Delta y \quad (I.30)$$

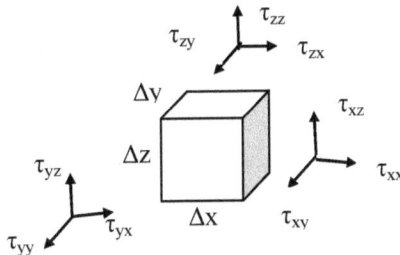

Similarly, the net force due to viscous stress on $\Delta x \Delta y$, in the x-direction, is

$$F_{net\,viscous\,stress} = \tau_{zx2}.\Delta x.\Delta y - \tau_{zx1}.\Delta x.\Delta y$$

$$= \left(\tau_{zx0} + \frac{\Delta z}{2} \frac{\partial \tau_{zx}}{\partial z} - \tau_{zx0} + \frac{\Delta z}{2} \frac{\partial \tau_{zx}}{\partial z} \right).\Delta x.\Delta y$$

$$= \frac{\partial \tau_{zx}}{\partial z} \Delta z.\Delta x.\Delta y \qquad (I.31)$$

and in the $\Delta z \Delta x$ plane,

$$F_{net\,viscous\,stress} = \tau_{yx2}.\Delta z.\Delta x - \tau_{yx1}.\Delta z.\Delta x$$

$$= \left(\tau_{yx0} + \frac{\Delta y}{2} \frac{\partial \tau_{yx}}{\partial y} - \tau_{yx0} + \frac{\Delta y}{2} \frac{\partial \tau_{yx}}{\partial y} \right).\Delta z.\Delta x$$

$$= \frac{\partial \tau_{yx}}{\partial y} \Delta y.\Delta z.\Delta x \qquad (I.32)$$

We can sum up these net forces in the x-direction from equations (I.29), (I.30), (I.31) and (I.32) as

$$F_{total\,net\,force} = \left(-\frac{\partial P_x}{\partial x} + \frac{\partial \tau_{xx}}{\partial x} + \frac{\partial \tau_{yx}}{\partial y} + \frac{\partial \tau_{zx}}{\partial z} \right) \Delta z.\Delta x.\Delta y \qquad (I.33)$$

Similarly for the y direction

$$F_{total\,net\,force} = \left(-\frac{\partial P_y}{\partial y} + \frac{\partial \tau_{xy}}{\partial x} + \frac{\partial \tau_{yy}}{\partial y} + \frac{\partial \tau_{zy}}{\partial z} \right) \Delta x.\Delta y.\Delta z \qquad (I.34)$$

And the z-direction

$$F_{total\,net\,force} = \left(-\frac{\partial P_z}{\partial z} + \frac{\partial \tau_{xz}}{\partial x} + \frac{\partial \tau_{yz}}{\partial y} + \frac{\partial \tau_{zz}}{\partial z} \right) \Delta x.\Delta y.\Delta z \qquad (I.35)$$

Another force acting on the fluid element is the body force, such as gravity, and it has components in the x, y, and z directions, designated by F_x, F_y, F_z, as force per unit mass. Usually, F_x and F_y are each zero. The gravity or body force, F_z, is given by $\Delta x.\Delta y.\Delta z.\rho.g$.

$$F_z = \rho.\Delta x.\Delta y.\Delta z.g \qquad (I.36)$$

The acceleration of the fluid element can be determined if we recall that its velocities, u, v, w, in the x, y, and z directions, are, each, functions of x, y, z, and t, the time. That is:

$$u = u(x, y, z, t)$$
$$v = v(x, y, z, t) \qquad (I.37)$$
$$w = w(x, y, z, t)$$

Thus, for u, for example

$$du = \frac{\partial u}{\partial t} dt + \frac{\partial u}{\partial x} dx + \frac{\partial u}{\partial y} dy + \frac{\partial u}{\partial z} dz \qquad (I.38)$$

and since

$$dx = u\,dt; \quad dy = v\,dt; \quad dz; \quad dz = w\,dt \qquad (I.39)$$

equation (I.38) becomes

$$du = \frac{\partial u}{\partial t} dt + u\frac{\partial u}{\partial x} dt + v\frac{\partial u}{\partial y} dt + w\frac{\partial u}{\partial z} dt \qquad (I.40)$$

The acceleration of fluid element in the x- direction, thus, becomes

$$\frac{du}{dt} = \frac{\partial u}{\partial t} + u\frac{\partial u}{\partial x} + v\frac{\partial u}{\partial y} + w\frac{\partial u}{\partial z} = \frac{Du}{Dt} \qquad (1.41)$$

Similarly for v in the y- direction:

$$\frac{Dv}{Dt} = \frac{\partial v}{\partial t} + u\frac{\partial v}{\partial x} + v\frac{\partial v}{\partial y} + w\frac{\partial v}{\partial z} \qquad (1.42)$$

And for w in the z - direction:

$$\frac{Dw}{Dt} = \frac{\partial w}{\partial t} + u\frac{\partial w}{\partial x} + v\frac{\partial w}{\partial y} + w\frac{\partial w}{\partial z} \qquad (1.43)$$

Thus, from equations (I.33), (I.34), (I.35), the body forces per unit mass, F_x, F_y, F_z, the acceleration of the fluid element described by equations (I.41), (I.42), (I.43) and Newton's second law of motion expressed in equation (I.1), we obtain the equation for fluid motion, called the Cauchy's equation of motion, as

In the x- direction

$$\left(-\frac{\partial P_x}{\partial x} + \frac{\partial \tau_{xx}}{\partial x} + \frac{\partial \tau_{yx}}{\partial y} + \frac{\partial \tau_{zx}}{\partial z}\right)\Delta z.\Delta x.\Delta y = \rho\,\Delta z.\Delta x.\Delta y\frac{Du}{Dt} \qquad (I.44)$$

which simplifies to

$$\frac{Du}{Dt} = \frac{1}{\rho}\left(-\frac{\partial P_x}{\partial x} + \frac{\partial \tau_{xx}}{\partial x} + \frac{\partial \tau_{yx}}{\partial y} + \frac{\partial \tau_{zx}}{\partial z}\right) \qquad (I.45)$$

Similar equations are obtained for the y-direction

$$\frac{Dv}{Dt} = \frac{1}{\rho}\left(-\frac{\partial P_y}{\partial y} + \frac{\partial \tau_{xy}}{\partial x} + \frac{\partial \tau_{yy}}{\partial y} + \frac{\partial \tau_{zy}}{\partial z}\right) \qquad (I.46)$$

and for the z-direction

$$\frac{Dw}{Dt} = \rho g + \frac{1}{\rho}\left(-\frac{\partial P_z}{\partial z} + \frac{\partial \tau_{xz}}{\partial x} + \frac{\partial \tau_{yz}}{\partial y} + \frac{\partial \tau_{zz}}{\partial z} \right) \qquad (I.47)$$

Constitutive Relationships

The mechanical behaviour of every substance can be described in terms of stress and strain. The relationship between these two is known as their constitutive relationship. For rigid solids, the constitutive relationship is between the stress tensor and the strain tensor while for fluids it is between the stress tensor and the strain rate tensor.

For the avoidance of doubt, Wikipedia (2009) helps us recall that in continuum mechanics, stress is a measure of the average force per unit area of a surface within a deformable body on which internal forces act. The Cauchy stress tensor is used for stress analysis of material bodies experiencing small deformations. For large deformations, also called finite deformations, other measures of stress are required, such as the first and second Piola-Kirchhoff stress tensors, the Biot stress tensor, and the Kirchhoff stress tensor.

Strain, also known as a deformation, and in continuum mechanics, is the change in the metric properties of a continuous body **B** in the displacement from an initial placement $\kappa_0(\mathbf{B})$ to a final placement $\kappa(\mathbf{B})$. A change in metric properties means that a curve drawn in the initial body placement changes its length when displaced to a curve in the final placement. If all the curves do not change length, a rigid body displacement is said to have occurred.

Strain is also a geometrical measure of deformation and represents the relative displacement between particles in a material body. It measures how much a given displacement differs locally from a rigid-body displacement. It also defines the amount of stretch or compression along a material line in elements or fibers, the *normal strain*, and the amount of distortion associated with the sliding of plane layers over each other, the *shear strain*, within a deforming body. Strain is a dimensionless quantity, which can be expressed as a decimal fraction, a percentage or in parts-per notation and can be applied to elongation, shortening, volume changes or angular distortion. The many types of strain are explained briefly below.
Engineering Strain

The **Cauchy strain** or **engineering strain** is expressed as the ratio of total deformation to the initial dimension of the material body in which the forces are being applied. The *engineering normal strain* or *engineering extensional strain, e,* of a material line element or fiber, axially loaded, is expressed as the change in length ΔL per unit of the original length L of the line element or fibers. Thus, we have

$$e = \frac{\Delta L}{L} = \frac{L_2 - L_1}{L_1} \qquad (1.48)$$

where *e* is the *engineering normal strain*, L_1 is the original length of the fiber and L_2 is the final length of the fiber. The normal strain is positive if the material fibers are stretched and negative if they are compressed.

The *engineering shear strain* is defined as the change in the angle between two material line elements initially perpendicular to each other in the undeformed or initial configuration.

Stretch Ratio

The **stretch ratio** or **extension ratio, λ,** is a measure of the extensional or normal strain of a differential line element, which can be defined at either the undeformed configuration or the deformed configuration. It is defined as the ratio between the final length L_2 and the initial length L_1 of the material line.

$$\lambda = \frac{L_2}{L_1} \qquad (1.49)$$

The extension ratio is related to the engineering strain by

$$e = \frac{L_2 - L_1}{L_1} = \lambda - 1 \qquad (1.50)$$

This equation implies that, when the stretch is equal to unity, the normal strain is zero, so that there is no deformation.

The stretch ratio is used in the analysis of materials that exhibit large deformations, such as elastometers, which can sustain stretch ratios of 3 or 4 before they fail. On the other hand, traditional engineering materials, such as concrete or steel, fail at much lower stretch ratios.
True Strain

The **logarithmic strain, ε,** is also called *natural strain, true strain* or *Hencky strain*. Considering an incremental strain (Ludwik)

$$\delta\varepsilon = \frac{\delta L}{L} \qquad (1.51)$$

the logarithmic strain is obtained by integrating this incremental strain:

$$\int_0^\varepsilon \delta\varepsilon = \int_{L_1}^{L_2} \frac{\delta L}{L} \quad or \quad \varepsilon = \ln\left(\frac{L_2}{L_1}\right) = \ln\lambda = \ln(1+e) = e - \frac{e^2}{2} + \frac{e^3}{3} - \ldots (1.52)$$

where e is the engineering strain. The logarithmic strain provides the correct measure of the final strain when deformation takes place in a series of increments, taking into account the influence of the strain path.

Green Strain

The **Green strain** is defined as

$$\varepsilon_G = \frac{1}{2}\left(\frac{L_2^2 - L_1^2}{L_1^2}\right) = \frac{1}{2}\left(\lambda^2 - 1\right) \qquad (1.53)$$

Almansi Strain

The **Euler-Almansi strain** is defined as

$$\varepsilon_E = \frac{1}{2}\left(\frac{L_2^2 - L_1^2}{L_2^2}\right) = \frac{1}{2}\left(1 - \frac{1}{\lambda^2}\right) \qquad (1.54)$$

Normal Strain

The normal strain in the x-direction of a rectangular element is defined by

$$\varepsilon = \frac{extension}{original length} = \frac{length(ab) - length(AB)}{length(AB)} = \frac{\partial u_x}{\partial x} \qquad (1.55)$$

Similarly, the normal strain, in the y- and z-directions, become

$$\varepsilon_y = \frac{\partial u_y}{\partial y} \quad and \quad \varepsilon_z = \frac{\partial u_z}{\partial z} \qquad (1.56)$$

Consider a two-dimensional infinitesimal rectangular material element with dimensions **dx** and **dy**, which after deformation, takes the form of a rhombus (Wikipedia, 2009).

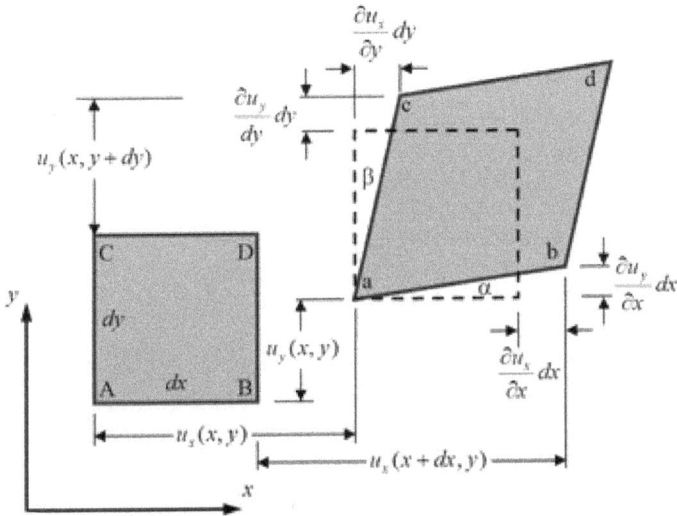

From the geometry of the figure we have

$$lengthof\ AB\ in\ rectangular element = d\,x \qquad (I.57)$$

and

$$lengthof\ ab\ in\ rhombus = \sqrt{\left(d\,x + \frac{\partial u_x}{\partial x}d\,x\right)^2 + \left(\frac{\partial u_y}{\partial x}d\,x\right)^2}$$

$$= d\,x\sqrt{1 + 2\frac{\partial u_x}{\partial x} + \left(\frac{\partial u_x}{\partial x}\right)^2 + \left(\frac{\partial u_y}{\partial x}\right)^2} \qquad (I.58)$$

For very small displacement gradients, the squares of the derivatives are negligible and we have

$$lengthof\ ab\ in\ rhombus \approx d\,x + \frac{\partial u_x}{\partial x}d\,x \qquad (I.59)$$

Shear Strain

The **engineering shear strain, γ,** is defined as the change in angle between two originally orthogonal material lines. In the figure above, the engineering shear strain (γ_{xy}) is the change in angle between lines AC and AB. Therefore,

$$\gamma_{xy} = \alpha + \beta \qquad (I.60)$$

From the geometry of the figure, we have

$$\tan\alpha = \frac{\dfrac{\partial u_y}{\partial x}dx}{dx + \dfrac{\partial u_x}{\partial x}dx} = \frac{\dfrac{\partial u_y}{\partial x}}{1 + \dfrac{\partial u_x}{\partial x}} \qquad (I.61)$$

$$\tan\beta = \frac{\dfrac{\partial u_x}{\partial y}dy}{dy + \dfrac{\partial u_y}{\partial y}dy} = \frac{\dfrac{\partial u_x}{\partial y}}{1 + \dfrac{\partial u_y}{\partial y}} \qquad (I.62)$$

For small displacement gradients we have

$$\frac{\partial u_x}{\partial x} \ll 1 \quad and \quad \frac{\partial u_y}{\partial y} \ll 1 \qquad (I.63)$$

For small rotations, that is when α and β are each less than 1, we have

$$\tan\alpha \approx \alpha \quad and \quad \tan\beta \approx \beta \qquad (I.64)$$

Therefore

$$\alpha \approx \frac{\partial u_y}{\partial x} \quad and \quad \beta \approx \frac{\partial u_x}{\partial y} \qquad (I.65)$$

Thus

$$\gamma_{xy} = \alpha + \beta = \frac{\partial u_y}{\partial x} + \frac{\partial u_x}{\partial y} \qquad (I.66)$$

By interchanging x and y and u_x and u_y, it can be shown that $\gamma_{xy} = \gamma_{yx}$. Similarly, for the y-z and x-z planes, we have

$$\gamma_{yz} = \gamma_{zy} = \frac{\partial u_y}{\partial z} + \frac{\partial u_z}{\partial y}, \quad \gamma_{zx} = \gamma_{xz} = \frac{\partial u_z}{\partial x} + \frac{\partial u_x}{\partial z} \qquad (I.67)$$

The tensorial shear strain components of the infinitesimal strain tensor can then be expressed using the engineering strain definition, γ, as

$$\varepsilon_{ij} = \begin{pmatrix} \varepsilon_{xx} & \varepsilon_{xy} & \varepsilon_{xz} \\ \varepsilon_{yx} & \varepsilon_{yy} & \varepsilon_{yz} \\ \varepsilon_{zx} & \varepsilon_{zy} & \varepsilon_{zz} \end{pmatrix} = \begin{pmatrix} \varepsilon_{xx} & \dfrac{\gamma_{xy}}{2} & \dfrac{\gamma_{xz}}{2} \\ \dfrac{\gamma_{yx}}{2} & \varepsilon_{yy} & \dfrac{\gamma_{yz}}{2} \\ \dfrac{\gamma_{zx}}{2} & \dfrac{\gamma_{zy}}{2} & \varepsilon_{zz} \end{pmatrix} \qquad (I.68)$$

From (1.66), (1.67) and (1.68), we can see that, at the centre of the parallelpiped, if for simplicity we replace u_x, u_y and u_z by u, v and w, respectively,

$$\varepsilon_{ij} = \begin{pmatrix} \dfrac{\partial u}{\partial x} & \dfrac{1}{2}\left(\dfrac{\partial u}{\partial y}+\dfrac{\partial v}{\partial x}\right) & \dfrac{1}{2}\left(\dfrac{\partial u}{\partial z}+\dfrac{\partial w}{\partial x}\right) \\[2mm] \dfrac{1}{2}\left(\dfrac{\partial u}{\partial y}+\dfrac{\partial v}{\partial x}\right) & \dfrac{\partial v}{\partial y} & \dfrac{1}{2}\left(\dfrac{\partial v}{\partial z}+\dfrac{\partial w}{\partial y}\right) \\[2mm] \dfrac{1}{2}\left(\dfrac{\partial u}{\partial z}+\dfrac{\partial w}{\partial x}\right) & \dfrac{1}{2}\left(\dfrac{\partial v}{\partial z}+\dfrac{\partial w}{\partial y}\right) & \dfrac{\partial w}{\partial z} \end{pmatrix} \qquad (I.69)$$

For a Newtonian fluid, the shear stress is linearly proportional to angular deformation. That is

$$\tau_{ij} = 2\mu\varepsilon_{ij} \qquad (I.70)$$

The constant of proportionality is known as the Newtonian viscosity. This enables us to express the shear stresses in a Newtonian fluid element as

$$\tau_{xy} = \tau_{yx} = \mu\left(\frac{\partial v}{\partial x}+\frac{\partial u}{\partial y}\right) \qquad (I.71)$$

$$\tau_{yz} = \tau_{zy} = \mu\left(\frac{\partial w}{\partial y}+\frac{\partial v}{\partial z}\right) \qquad (I.72)$$

$$\tau_{zx} = \tau_{xz} = \mu\left(\frac{\partial u}{\partial z}+\frac{\partial w}{\partial x}\right) \qquad (I.73)$$

When we substitute these in equations (1.45), (1.46) and (1.47) we get, in the x-direction

$$\frac{Du}{Dt} = -\frac{1}{\rho}\frac{\partial P_x}{\partial x} + \frac{1}{\rho}\left(\frac{\partial \tau_{xx}}{\partial x}+\frac{\partial \tau_{yx}}{\partial y}+\frac{\partial \tau_{zx}}{\partial z}\right) \qquad from\ (1.45)$$

$$= -\frac{1}{\rho}\frac{\partial P_x}{\partial x} + \frac{1}{\rho}\left(2\mu\frac{\partial}{\partial x}\left(\frac{\partial u}{\partial x}\right)+\mu\frac{\partial}{\partial y}\left(\frac{\partial v}{\partial x}+\frac{\partial u}{\partial y}\right)+\frac{\partial}{\partial z}\mu\left(\frac{\partial u}{\partial z}+\frac{\partial w}{\partial x}\right)\right)$$

$$= -\frac{1}{\rho}\frac{\partial P_x}{\partial x} + \frac{\mu}{\rho}\left(2\frac{\partial^2 u}{\partial x^2}+\frac{\partial^2 v}{\partial y\partial x}+\frac{\partial^2 u}{\partial y^2}+\frac{\partial^2 u}{\partial z^2}+\frac{\partial^2 w}{\partial z\partial x}\right)$$

$$= -\frac{1}{\rho}\frac{\partial P_x}{\partial x} + \frac{\mu}{\rho}\left(\frac{\partial^2 u}{\partial x^2}+\frac{\partial^2 u}{\partial y^2}+\frac{\partial^2 u}{\partial z^2}+\frac{\partial^2 u}{\partial x^2}+\frac{\partial^2 v}{\partial y\partial x}+\frac{\partial^2 w}{\partial z\partial x}\right)$$

$$= -\frac{1}{\rho}\frac{\partial P_x}{\partial x} + \frac{\mu}{\rho}\left(\frac{\partial^2 u}{\partial x^2}+\frac{\partial^2 u}{\partial y^2}+\frac{\partial^2 u}{\partial z^2}\right)+\frac{\mu}{\rho}\frac{\partial}{\partial x}\left(\frac{\partial u}{\partial x}+\frac{\partial v}{\partial y}+\frac{\partial w}{\partial z}\right) \qquad (I.74)$$

For incompressible flow

$$\left(\frac{\partial u}{\partial x}+\frac{\partial v}{\partial y}+\frac{\partial w}{\partial z}\right) = 0 \qquad (I.75)$$

201

So that equation (1.74) becomes

$$\frac{Du}{Dt} = -\frac{1}{\rho}\frac{\partial P_x}{\partial x} + \frac{\mu}{\rho}\left(\frac{\partial^2 u}{\partial x^2} + \frac{\partial^2 u}{\partial y^2} + \frac{\partial^2 u}{\partial z^2}\right) \qquad (I.76)$$

Similar equations are obtained for the y-direction

$$\frac{Dv}{Dt} = -\frac{1}{\rho}\frac{\partial P_y}{\partial y} + \frac{\mu}{\rho}\left(\frac{\partial^2 v}{\partial x^2} + \frac{\partial^2 v}{\partial y^2} + \frac{\partial^2 v}{\partial z^2}\right) \qquad (I.77)$$

and for the z-direction

$$\frac{Dw}{Dt} = g - \frac{1}{\rho}\frac{\partial P_z}{\partial z} + \frac{\mu}{\rho}\left(\frac{\partial^2 w}{\partial x^2} + \frac{\partial^2 w}{\partial y^2} + \frac{\partial^2 w}{\partial z^2}\right) \qquad (I.78)$$

Equations (I.76), (I.77) and (I.78) can, again, be expressed in concise vector notation as in equation (I.26)

$$\rho\frac{Dv}{Dt} = -\nabla p + \mu\nabla^2 v + \rho g \qquad (I.26)$$

where v is, now, a general velocity vector.

APPENDIX II
The Navier Stokes Equations in Other Co-ordinate Systems

Table II.1: The Navier Stokes Equations in Cartesian Co-ordinates

Continuity Equation

$$\frac{\partial u}{\partial x} + \frac{\partial v}{\partial y} + \frac{\partial w}{\partial z} = 0 \qquad (II.1)$$

Momentum Equation in x-direction

$$\frac{\partial u}{\partial t} + u\frac{\partial u}{\partial x} + v\frac{\partial u}{\partial y} + w\frac{\partial u}{\partial z} = -\frac{1}{\rho}\frac{\partial P_x}{\partial x} + \frac{\mu}{\rho}\left(\frac{\partial^2 u}{\partial x^2} + \frac{\partial^2 u}{\partial y^2} + \frac{\partial^2 u}{\partial z^2}\right) \qquad (II.2)$$

Momentum Equation in y-direction

$$\frac{\partial v}{\partial t} + u\frac{\partial v}{\partial x} + v\frac{\partial v}{\partial y} + w\frac{\partial v}{\partial z} = -\frac{1}{\rho}\frac{\partial P_y}{\partial y} + \frac{\mu}{\rho}\left(\frac{\partial^2 v}{\partial x^2} + \frac{\partial^2 v}{\partial y^2} + \frac{\partial^2 v}{\partial z^2}\right) \qquad (II.3)$$

Momentum Equation in z-direction

$$\frac{\partial w}{\partial t} + u\frac{\partial w}{\partial x} + v\frac{\partial w}{\partial y} + w\frac{\partial w}{\partial z} = g - \frac{1}{\rho}\frac{\partial P_z}{\partial z} + \frac{\mu}{\rho}\left(\frac{\partial^2 w}{\partial x^2} + \frac{\partial^2 w}{\partial y^2} + \frac{\partial^2 w}{\partial z^2}\right) \qquad (II.4)$$

Energy Equation

$$\frac{\partial T}{\partial t} + u\frac{\partial T}{\partial x} + v\frac{\partial T}{\partial y} + w\frac{\partial T}{\partial z} = \frac{k}{\rho Cp}\left(\frac{\partial^2 T}{\partial x^2} + \frac{\partial^2 T}{\partial y^2} + \frac{\partial^2 T}{\partial z^2}\right) \qquad (II.5)$$

where
- u = velocity in the x-direction
- v = velocity in the y-direction
- w = velocity in the z-direction
- P = pressure
- T = temperature
- k = thermal conductivity
- ρ = density
- Cp = specific heat capacity

Table II.2: The Navier Stokes Equations in Cylindrical Co-ordinates

Continuity Equation

$$\frac{1}{r}\frac{\partial (r v_r)}{\partial r}+\frac{1}{r}\frac{\partial v_\theta}{\partial \theta}+\frac{\partial w}{\partial z}=0 \qquad (II.6)$$

Momentum Equation in r-direction

$$\frac{\partial v_r}{\partial t}+v_r\frac{\partial v_r}{\partial r}+\frac{v_\theta}{r}\frac{\partial v_r}{\partial \theta}-\frac{v_\theta^2}{r}+w\frac{\partial v_r}{\partial z}$$

$$=g_r-\frac{1}{\rho}\frac{\partial P}{\partial r}+\frac{\mu}{\rho}\left[\frac{\partial}{\partial r}\left(\frac{1}{r}\frac{\partial (r v_r)}{\partial r}\right)+\frac{1}{r^2}\frac{\partial^2 v_r}{\partial \theta^2}-\frac{2}{r^2}\frac{\partial v_\theta}{\partial \theta}+\frac{\partial^2 v_r}{\partial z^2}\right] \qquad (II.7)$$

Momentum Equation in θ-direction

$$\frac{\partial v_\theta}{\partial t}+v_r\frac{\partial v_\theta}{\partial r}+\frac{v_\theta}{r}\frac{\partial v_\theta}{\partial \theta}-\frac{v_r\cdot v_\theta}{r}+w\frac{\partial v_\theta}{\partial z}$$

$$=g_\theta-\frac{1}{\rho r}\frac{\partial P}{\partial \theta}+\frac{\mu}{\rho}\left[\frac{\partial}{\partial r}\left(\frac{1}{r}\frac{\partial (r v_\theta)}{\partial r}\right)+\frac{1}{r^2}\frac{\partial^2 v_\theta}{\partial \theta^2}+\frac{2}{r^2}\frac{\partial v_r}{\partial \theta}+\frac{\partial^2 v_\theta}{\partial z^2}\right] \qquad (II.8)$$

Momentum Equation in z-direction

$$\frac{\partial w}{\partial t}+v_r\frac{\partial w}{\partial r}+\frac{v_\theta}{r}\frac{\partial w}{\partial \theta}+w\frac{\partial w}{\partial z}$$

$$=g_z-\frac{1}{\rho}\frac{\partial P}{\partial z}+\frac{\mu}{\rho}\left[\frac{1}{r}\frac{\partial}{\partial r}\left(r\frac{\partial w}{\partial r}\right)+\frac{1}{r^2}\frac{\partial^2 w}{\partial \theta^2}+\frac{\partial^2 w}{\partial z^2}\right] \qquad (II.9)$$

Energy Equation

$$\frac{\partial T}{\partial t}+v_r\frac{\partial T}{\partial r}+\frac{v_\theta}{r}\frac{\partial T}{\partial \theta}+w\frac{\partial T}{\partial z}=\frac{k}{\rho Cp}\left[\frac{1}{r}\frac{\partial}{\partial r}\left(r\frac{\partial T}{\partial r}\right)+\frac{1}{r^2}\frac{\partial^2 T}{\partial \theta^2}+\frac{\partial^2 T}{\partial z^2}\right] \qquad (II.10)$$

where

v_r = velocity in the r-direction
v_θ = velocity in the θ-direction
w = velocity in the z-direction
P = pressure
T = temperature
k = thermal conductivity
ρ = density
Cp = specific heat capacity

Table II.3: The Navier Stokes Equations in Spherical Co-ordinates

Continuity Equation

$$\frac{1}{r^2}\frac{\partial(r^2 v_r)}{\partial r} + \frac{1}{r\sin\theta}\frac{\partial(v_\theta \sin\theta)}{\partial\theta} + \frac{1}{r\sin\theta}\frac{\partial v_\varphi}{\partial\varphi} = 0 \qquad (II.11)$$

Momentum Equation in r-direction

$$\frac{\partial v_r}{\partial t} + v_r\frac{\partial v_r}{\partial r} + \frac{v_\theta}{r}\frac{\partial v_r}{\partial\theta} + \frac{v_\varphi}{r\sin\theta}\frac{\partial v_r}{\partial\varphi} - \frac{v_\theta^2 + v_\varphi^2}{r}$$

$$= g_r - \frac{1}{\rho}\frac{\partial P}{\partial r} + \frac{\mu}{\rho}\left[\nabla^2 v_r - \frac{2v_r}{r^2} - \frac{2}{r^2}\frac{\partial v_\theta}{\partial\theta} - \frac{2}{r^2}v_\theta\cot\theta - \frac{2}{r^2\sin\theta}\frac{\partial v_\varphi}{\partial\varphi}\right] \qquad (II.12)$$

Momentum Equation in θ-direction

$$\frac{\partial v_\theta}{\partial t} + v_r\frac{\partial v_\theta}{\partial r} + \frac{v_\theta}{r}\frac{\partial v_\theta}{\partial\theta} + \frac{v_r\cdot v_\theta}{r} + \frac{v_\varphi}{r\sin\theta}\frac{\partial v_\theta}{\partial\varphi} - \frac{v_\varphi^2\cot\theta}{r}$$

$$= g_\theta - \frac{1}{\rho}\frac{1}{r}\frac{\partial P}{\partial\theta} + \frac{\mu}{\rho}\left[\nabla^2 v_\theta + \frac{2}{r^2}\frac{\partial v_r}{\partial\theta} - \frac{v_\theta}{r^2\sin^2\theta} - \frac{2\cos\theta}{r^2\sin^2\theta}\frac{\partial v_\varphi}{\partial\varphi}\right] \qquad (II.13)$$

Momentum Equation in φ-direction

$$\frac{\partial v_\varphi}{\partial t} + v_r\frac{\partial v_\varphi}{\partial r} + v_\theta\frac{\partial v_\varphi}{\partial\theta} + \frac{v_r\cdot v_\varphi}{r} + \frac{v_\varphi}{r\sin\theta}\frac{\partial v_\varphi}{\partial\varphi} + \frac{v_\theta\cdot v_\varphi}{r}\cot\theta = g_\varphi$$

$$- \frac{1}{\rho}\frac{1}{r\sin\theta}\frac{\partial P}{\partial\varphi} + \frac{\mu}{\rho}\left[\nabla^2 v_\varphi - \frac{v_\varphi}{r^2\sin^2\theta} + \frac{2}{r^2\sin\theta}\frac{\partial v_r}{\partial\varphi} + \frac{2\cos\theta}{r^2\sin^2\theta}\frac{\partial v_\varphi}{\partial\varphi}\right] (II.14)$$

Energy Equation

$$\frac{\partial T}{\partial t} + v_r\frac{\partial T}{\partial r} + \frac{v_\theta}{r}\frac{\partial T}{\partial\theta} + \frac{v_\varphi}{r\sin\theta}\frac{\partial T}{\partial\varphi}$$

$$= \frac{k}{\rho Cp}\left[\frac{1}{r}\frac{\partial}{\partial r}\left(r^2\frac{\partial T}{\partial r}\right) + \frac{1}{r^2\sin\theta}\frac{\partial}{\partial\theta}\left(\sin\theta\frac{\partial T}{\partial\theta}\right) + \frac{1}{r^2\sin^2\theta}\frac{\partial^2 T}{\partial\varphi^2}\right] \qquad (II.15)$$

where

$$\nabla^2 = \frac{1}{r^2}\frac{\partial}{\partial r}\left(r^2\frac{\partial}{\partial r}\right) + \frac{1}{r^2\sin\theta}\frac{\partial}{\partial\theta}\left(\sin\theta\frac{\partial}{\partial\theta}\right) + \frac{1}{r^2\sin^2\theta}\frac{\partial^2}{\partial\varphi^2} \qquad (II.16)$$

and

$$v_r = \quad \text{velocity in the r-direction}$$

$v_\theta =$ velocity in the θ-direction

$v_\varphi =$ velocity in the φ-direction

$P =$ pressure

$T =$ temperature

$k =$ thermal conductivity

$\rho =$ density

$Cp =$ specific heat capacity

APPENDIX III
Analytical Solutions of the Navier-Stokes Equations

As the scientific and engineering community became increasingly confident in mathematical procedures and analyses, and faced with demands for mathematical solutions to the technical problems of the emerging commercial industries, a great deal of effort was put into solving the Navier-Stokes equations analytically. The results of these efforts were what we can, now, classify as exact and approximate solutions and varied from industry to industry. In the manufacturing and chemical industry, most of these solutions sought to derive mathematical expressions for estimating fluid friction or heat transfer coefficients in the boundary layer regions of fluid flow.

Exact Solutions

Exact solutions are not really as exact as the name would suggest. They are not exact because some simplifying assumptions (usually, order of magnitude assumptions) are made so that certain terms are removed from the equations. The solution is exact, however, because the equation that is left is solved exactly.

The simplest example is the solution of the Navier Stokes equation for a large flat plate, initially, at rest but suddenly accelerated from rest and moved in its own plane with velocity u_o. It is desired to find the velocity and temperature distribution of fluid in the vicinity of this plate.

To solve this problem, consider the momentum equation, in vector form

$$\rho \frac{Du}{Dt} = -\nabla P + \mu \nabla^2 u \qquad (III.1)$$

In scalar form and in the two dimensions, x and y, this becomes, in the:
x-direction

$$\rho \left(\frac{\partial u}{\partial t} + u \frac{\partial u}{\partial x} + v \frac{\partial u}{\partial y} \right) = -\frac{\partial P}{\partial x} + \mu \left(\frac{\partial^2 u}{\partial x^2} + \frac{\partial^2 u}{\partial y^2} \right) \qquad (III.2)$$

y-direction

$$\rho \left(\frac{\partial v}{\partial t} + u \frac{\partial v}{\partial x} + v \frac{\partial v}{\partial y} \right) = -\frac{\partial P}{\partial y} + \mu \left(\frac{\partial^2 v}{\partial x^2} + \frac{\partial^2 v}{\partial y^2} \right) \qquad (III.3)$$

The continuity equation is

207

$$\left(\frac{\partial u}{\partial x} + \frac{\partial v}{\partial y}\right) = 0 \qquad (III.4)$$

The energy equation is

$$\frac{\partial T}{\partial t} + u\frac{\partial T}{\partial x} + v\frac{\partial T}{\partial y} = \frac{k}{\rho Cp} \cdot \left(\frac{\partial^2 T}{\partial x^2} + \frac{\partial^2 T}{\partial y^2}\right) \qquad (III.5)$$

The initial and boundary conditions, as defined by the problem, are

$t \leq 0$ **u** = 0 all y

$t > 0$ **u** = **u₀** at y = 0   (III.6)

$t > 0$ **u** → **0** as y → ∞

To solve equations (III.2) to (III.5) subject to (III.6), simplifying assumptions need to be made. These are as follows;

<u>Simplifying assumptions</u>

We assume, from the statement of the problem, that only the boundary layer is involved. This makes $v = 0$ for all x and y. Hence from continuity

$$\frac{\partial u}{\partial x} = 0 \qquad (III.7)$$

There is no pressure gradient in the x direction so that

$$\frac{\partial P}{\partial x} = 0 \qquad (III.8)$$

Because the boundary layer is very thin

$$\frac{\partial P}{\partial y} = 0 \qquad (III.9)$$

All these enable us to simplify the momentum equation to

$$\frac{\partial u}{\partial t} = \frac{\mu}{\rho} \cdot \frac{\partial^2 u}{\partial y^2} = v\frac{\partial^2 u}{\partial y^2} \qquad (III.10)$$

with the same boundary conditions as before, where $v = \frac{\mu}{\rho}$.

Equation (III.10) may be solved analytically using either a separation of variables, Fourier sine transform or Laplace transform, technique. The Laplace transform technique is used in this case because it is the simplest and also gives solutions applicable to short times as implied in the problem. This solution is

$$u(t, y) = u_o \, erfc\left(\frac{y}{2\sqrt{vt}}\right) \qquad (III.11)$$

208

In a similarly manner, the energy equation is reduced to

$$\frac{\partial T}{\partial t} = \frac{k}{\rho C p} \cdot \frac{\partial^2 T}{\partial y^2} = \alpha \frac{\partial^2 T}{\partial y^2} \qquad (III.12)$$

with the solution

$$T(t, y) = T_o \, erfc\left(\frac{y}{2\sqrt{\alpha t}}\right) \qquad (III.13)$$

These are the temperature and velocity distributions required in the problem.

If the boundary layer thickness, δ, is defined as that distance from the solid surface at which fluid velocity equals 99% of the undisturbed mainstream velocity or at which the temperature is 99% of that at the surface, as is shown below for velocity,

Definition of the Boundary Layer Thickness, δ

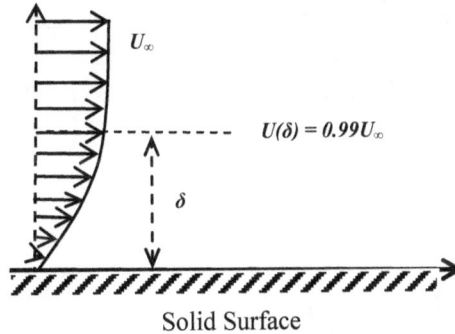

Solid Surface

then, at the boundary layer, represented by δ_H, for the hydrodynamic boundary layer and by δ_T, for the thermal boundary layer,

$$\frac{u(t, y)}{u_\infty} = 0.01 = erfc\left(\frac{\delta_H}{2\sqrt{vt}}\right) = \frac{T(t, y)}{T_\infty} = erfc\left(\frac{\delta_T}{2\sqrt{\alpha t}}\right) \qquad (III.14)$$

Since *erfc 2* $= 0.005 \approx 0.01$, we can see, from (III.14) that

$$\delta_H = 4.\sqrt{vt}; \quad \delta_T = 4.\sqrt{\alpha t}; \quad and \quad \frac{\delta_H}{\delta_T} = \sqrt{\frac{v}{\alpha}} = \sqrt{Pr} \qquad (III.15)$$

Equation (III.15) tells us that if

$$\begin{array}{lll} Pr = 1, \text{ then} & \delta_H = \delta_T & \\ Pr > 1, \text{ then} & \delta_H > \delta_T & (III.16) \\ Pr < 1, \text{ then} & \delta_H < \delta_T & \end{array}$$

Since $t = \dfrac{x}{u_\infty}$ where x is any distance travelled in time t at velocity u_∞

$$\delta_H = 4.\sqrt{v\frac{x}{u_\infty}} = 4.\sqrt{\frac{v x^2}{u_\infty x}} = \frac{4x}{\sqrt{Re_x}} \qquad (III.17)$$

$$\delta_T = 4.\sqrt{\alpha\frac{x}{u_\infty}} = 4.\sqrt{\frac{x^2 \mu k}{\rho u_\infty x Cp\mu}} = \frac{4x}{\sqrt{Re_x\,Pr}} \qquad (III.18)$$

It is usual, however, to define $\delta_H^* = \dfrac{\delta_H}{D_c}$ where D_c is a characteristic dimension. This gives, from equation (III.17),

$$\delta_H^* = \frac{\delta_H}{D_c} = 4.\sqrt{v\frac{x}{u_\infty D_c^2}} = 4.\sqrt{\frac{v}{u_\infty D_c}\cdot\frac{x}{D_c}} = \frac{4}{\sqrt{Re}}\cdot\left(\frac{x}{D_c}\right)^{\frac{1}{2}} \qquad (III.19)$$

The heat flux across the wall and the boundary layer is given by

$$q = -k\frac{\partial T}{\partial y} = h(T_w - T_\infty) \qquad (III.20)$$

$\dfrac{\partial T}{\partial y}$ may be obtained from (III.13) and substituted in (III.20) to get

$$q = -k\left(T_w - T_\infty\right)\frac{\partial}{\partial y}\left[erfc\left(\frac{y}{2\sqrt{\alpha t}}\right)\right]$$

$$= -k\left(T_w - T_\infty\right)\frac{\partial}{\partial y}\left[1 - \frac{2}{\pi}\int_0^u e^{-u^2}\,du\right] = \left[k\left(T_w - T_\infty\right).\frac{2}{\pi}.\frac{1}{2\sqrt{\alpha t}}.e^{-\frac{y^2}{4\alpha t}}\right]$$

$$= \frac{k\left(T_w - T_\infty\right)}{\sqrt{\pi\alpha t}} = \frac{\left(T_w - T_\infty\right)}{\sqrt{\dfrac{\pi\alpha x}{u_\infty}}} \qquad (III.21)$$

where $t = x/u_\infty$, T_w is temperature at the wall and T_∞ is temperature in the main body of the fluid.

Equation (III.21) may, also, be expressed in a more general form if we recall that the Nusselt number may be expressed as

$$Nu = \frac{q_\infty x}{k\,\Delta T} = \frac{k\left(T_w - T_\infty\right)}{\sqrt{\dfrac{\pi\alpha x}{u_\infty}}}.\frac{x}{k\left(T_w - T_\infty\right)} = x\sqrt{\frac{u_\infty}{\pi\alpha x}} = \sqrt{\frac{u_\infty x^2}{\pi\alpha x}.\frac{v}{v}}$$

$$= \sqrt{\frac{\nu}{\alpha} \cdot \frac{\rho u_\infty x}{\mu} \cdot \frac{1}{\pi}} = \sqrt{\frac{Pr.Re}{\pi}} = 0.564 Re^{\frac{1}{2}}.Pr^{\frac{1}{2}} \qquad (III.22)$$

Exact solutions begin to get more complicated when more complicated fluid flow and heat transfer configurations are to be analysed such as for example, fluid flow and heat transfer in laminar or turbulent flow along flat plates or inside circular pipes.

Approximate or Integral Solutions

Approximate methods are able to handle slightly more complex configurations better. The approach is based on the von Karman integral equation for momentum and heat transfer derived on the basis of a momentum and heat energy balance around the particular fluid flow or heat transfer configuration.

For heat transfer in laminar boundary flow, for example, consider the schematic below

Momentum Balance in a Boundary Layer

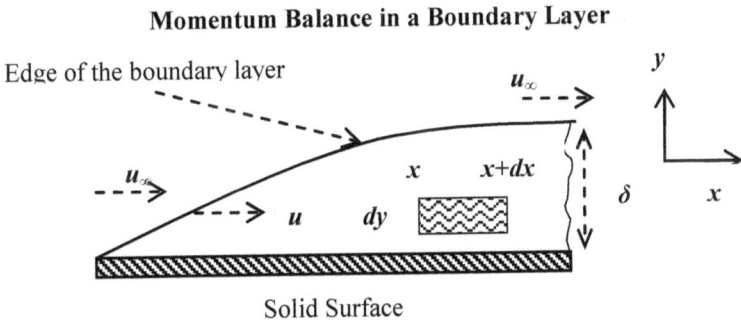

Solid Surface

Conservation of mass requires that, over the control volume, C_V, and control surface, C_S, defined by Δx and Δy and unit Δz,

$$\frac{\partial}{\partial t} \int_{C_V} \rho \, dV + \int_{C_S} \rho(u.n) \, dA = 0 \qquad III.23)$$

A force balance, according to Newton's second law of motion, gives

$$\sum F = \frac{\partial}{\partial t} \int_{C_V} u \rho \, dV + \int_{C_S} u \rho(u.n) \, dA \qquad (III.24)$$

where F is the momentum flux, u represents velocity, V represents volume, n is the normal to the surface and A is surface area.

At steady state $\qquad \frac{\partial}{\partial t} \int_{C_V} \rho \, dV = 0 \qquad III.25)$

211

and

$$\int_{C_S} \rho(u.n)dA = \int_0^\delta \rho u\, dy\big|_{x+\Delta x} - \int_0^\delta \rho u\, dy\big|_x - m_y.\Delta x = 0 \qquad (III.26)$$

where m_y is the rate of loss, or exchange of fluid, from the control volume in the y direction.

Equation (III.26) can be rearranged as

$$m_y = \frac{\int_0^\delta \rho u\, dy\big|_{x+\Delta x} - \int_0^\delta \rho u\, dy\big|_x}{\Delta x} \qquad (III.27)$$

which, in the limit as $\Delta x \to 0$, becomes

$$m_y = \frac{d}{dx}\int_0^\delta \rho u\, dy \qquad (III.28)$$

Since

$$\sum F = -\tau_{yx}.\Delta x + \int_0^\delta P\, dy\big|_x - \int_0^\delta P\, dy\big|_{x+\Delta x} \qquad (III.29)$$

where P is pressure in differential element and τ_{yx} is the shear stress on y along the x direction

$$-\tau_{yx}.\Delta x + \int_0^\delta P\, dy\big|_x - \int_0^\delta P\, dy\big|_{x+\Delta x} = \int_{C_S} u\rho(u.n)dA$$

$$= \int_0^\delta \rho u^2\, dy\big|_{x+\Delta x} - \int_0^\delta \rho u^2\, dy\big|_x - u_\infty m_y.\Delta x$$

$$= \int_0^\delta \rho u^2\, dy\big|_{x+\Delta x} - \int_0^\delta \rho u^2\, dy\big|_x - u_\infty.\Delta x.\frac{d}{dx}\int_0^\delta \rho u\, dy \qquad (III.30)$$

In the limit as $\Delta x \to 0$ and $\tau_{yx} \to \tau_o$ equation (III.30) reduces to

$$-\tau_o - \frac{d}{dx}\int_0^\delta P\, dy = \frac{d}{dx}\int_0^\delta \rho u^2\, dy - u_\infty \frac{d}{dx}\int_0^\delta \rho u\, dy \qquad (III.31)$$

where τ_o is the shear stress at the wall. Since P is not a function of y

$$\frac{d}{dx}\int_0^\delta P\, dy = \frac{dP}{dx}\int_0^\delta dy = \frac{y\, dP}{dx} \qquad (III.32)$$

By Bernoulli's equation.

$$\frac{dP}{dx} = -\rho u_\infty \frac{du_\infty}{dx} \qquad (III.33)$$

Combining equations (III.31), (III.32) and (III.33) from zero to the thickness of the boundary layer, δ, we get that

$$\tau_0 - y\rho u_\infty \frac{du_\infty}{dx} = \frac{d}{dx} \int_0^\delta \rho u (u_\infty - u) dy \qquad (III.34)$$

- the von Karman momentum integral equation.

In a similar manner to the analysis of momentum transfer in the boundary layer shown in the diagram above, the heat energy balance is given by

$$-k\Delta x \frac{\partial T}{\partial y}\bigg|_{y=0}$$

$$= \int_0^\delta \rho u Cp T dy\bigg|_{x+\Delta x} - \int_0^\delta \rho u Cp T dy\bigg|_x - \rho Cp\Delta x.\frac{d}{dx}\int_0^\delta u T_\infty dy$$

This gives the integral equation for heat transfer in the thermal boundary layer as

$$\frac{k}{\rho Cp}\frac{\partial T}{\partial y}\bigg|_{y=0} = \frac{d}{dx}\int_0^\delta u(T_\infty - T)dy \qquad (III.35)$$

We can, now, use equations (III.34) and (III.35) to evaluate momentum and heat transfer in the boundary layer.

Solution of the von Karman Integral Equations using only Velocity and Temperature Profiles

For example, if we assume a general velocity profile,

$$u = a + by + cy^2 + dy^3 \quad for \quad 0 \le y \le \delta \qquad (III.36a)$$

$$= u_\infty \qquad\qquad\qquad for \quad \delta \le y \qquad (III.36b)$$

where a, b, c, d are constants, with the boundary conditions that at

i $y = 0$ $u = 0$

ii $y = \delta$ $u \approx u_\infty$

iii $y = \delta$ $\dfrac{du}{dy} = 0$ (III.37)

iv $y = \delta$ $\dfrac{d^2 u}{dy^2} = 0$

we can determine the constants of equation (III.36) to obtain the velocity profile as

$$u = u_\infty \left(\frac{3y}{\delta} - \frac{3y^2}{\delta^2} + \frac{y^3}{\delta^3} \right) \quad for \quad 0 \le y \le \delta$$

$$= u_\infty \qquad\qquad for \quad y \ge \delta \qquad\qquad (III.38)$$

When we substitute (III.38) into the von-Karman momentum integral equation (III.34), we get, since u_∞ is independent of x,

$$\tau_o = \frac{d}{dx} \int_0^\delta \rho u_\infty \left(\frac{3y}{\delta} - \frac{3y^2}{\delta^2} + \frac{y^3}{\delta^3} \right) \left(u_\infty - u_\infty \left(\frac{3y}{\delta} - \frac{3y^2}{\delta^2} + \frac{y^3}{\delta^3} \right) \right) dy$$

$$= \frac{d}{dx} \int_0^\delta \rho u_\infty^2 \left(\frac{3y}{\delta} - \frac{3y^2}{\delta^2} + \frac{y^3}{\delta^3} \right) \left(1 - \left(\frac{3y}{\delta} - \frac{3y^2}{\delta^2} + \frac{y^3}{\delta^3} \right) \right) dy$$

$$= \frac{d}{dx} \int_0^\delta \rho u_\infty^2 \left[\left(\frac{3y}{\delta} - \frac{3y^2}{\delta^2} + \frac{y^3}{\delta^3} \right) - \left(\frac{3y}{\delta} - \frac{3y^2}{\delta^2} + \frac{y^3}{\delta^3} \right)\left(\frac{3y}{\delta} - \frac{3y^2}{\delta^2} + \frac{y^3}{\delta^3} \right) \right] dy$$

$$= \frac{d}{dx} \int_0^\delta \rho u_\infty^2 \left(\frac{3y}{\delta} - \frac{3y^2}{\delta^2} + \frac{y^3}{\delta^3} \right) dy$$

$$- \frac{d}{dx} \int_0^\delta \left(\frac{9y^2}{\delta^2} - \frac{9y^3}{\delta^3} + \frac{3y^4}{\delta^4} - \frac{9y^3}{\delta^3} + \frac{9y^4}{\delta^4} - \frac{3y^5}{\delta^5} + \frac{3y^4}{\delta^4} - \frac{3y^5}{\delta^5} + \frac{y^6}{\delta^6} \right) dy$$

$$= \frac{d}{dx} \int_0^\delta \rho u_\infty^2 \left(\frac{3y}{\delta} - \frac{12y^2}{\delta^2} + \frac{19y^3}{\delta^3} - \frac{15y^4}{\delta^4} + \frac{6y^5}{\delta^5} - \frac{y^6}{\delta^6} \right) dy$$

$$= \frac{d}{dx} \left[\rho u_\infty^2 \left(\frac{3}{2} \frac{\delta^2}{\delta} - \frac{4\delta^3}{\delta^2} + \frac{19\delta^4}{4\delta^3} - \frac{3\delta^5}{\delta^4} + \frac{\delta^6}{\delta^5} - \frac{\delta^7}{7\delta^6} \right) \right]$$

$$= \rho u_\infty^2 \frac{d\delta}{dx} \left(\frac{42 - 112 + 133 - 84 + 28 - 4}{28} \right) = \frac{3}{28} \rho u_\infty^2 \frac{d\delta}{dx} \qquad (III.39)$$

The shear stress at the wall τ_o is given, for a Newtonian fluid, by

$$\tau_o = \mu \frac{\partial u}{\partial y}\bigg|_{y=0} = \mu \frac{\partial}{\partial y} \left[u_\infty \left(\frac{3y}{\delta} - \frac{3y^2}{\delta^2} + \frac{y^3}{\delta^3} \right) \right]_{y=0} = \frac{3\mu u_\infty}{\delta} \qquad (III.40)$$

From (III.39) and (III.40) we get

$$\frac{3}{28} \rho u_\infty^2 \frac{d\delta}{dx} = \frac{3\mu u_\infty}{\delta} \quad or \quad \delta d\delta = \frac{28\mu}{\rho u_\infty} dx \qquad (III.41)$$

which, on integration and rearranging, gives

$$\frac{\delta}{x} = 7.483\left(\frac{\mu}{\rho u_\infty x}\right) = \frac{7.483}{\sqrt{Re_x}} \qquad (III.42)$$

We can see that the local coefficient of skin friction, C_{fx}, becomes

$$C_{fx} = \frac{\tau_o}{\rho u_\infty^2 \Big/ 2} = \frac{3\mu u_\infty}{\delta} \cdot \frac{2}{\rho u_\infty^2} = \frac{6\mu (\rho u_\infty x)^{\frac{1}{2}}}{7.483\rho u_\infty x \mu^{\frac{1}{2}}} = 0.802\left(\frac{\mu}{\rho u_\infty x}\right)^{\frac{1}{2}}$$

$$= 0.802 Re_x^{-\frac{1}{2}} \qquad (III.43)$$

and the mean coefficient of skin friction

$$C_{fL} = \frac{1}{L}\int_0^L C_{fx}\, dx = 1.604 Re_L^{-\frac{1}{2}} \qquad (III.44)$$

Note that these results are similar to those obtained by exact analysis. Similarly, we can assume a general temperature profile,

$$T - T_o = A + By + Cy^2 + Dy^3 \quad for \quad 0 \le y \le \delta \qquad (III.45a)$$

$$= T_\infty - T_o \qquad\qquad for \quad \delta \le y \qquad (III.45b)$$

where A, B, C, D are constants and T_o the temperature at the wall, with the boundary conditions that at

i $y = 0$ $T - T_o = 0$

ii $y = \delta$ $T - T_o = T_\infty - T_o$

iii $y = \delta$ $\dfrac{d(T - T_o)}{dy} = 0$ (III.46)

iv $y = \delta$ $\dfrac{d^2(T - T_o)}{dy^2} = 0$

In a similar manner to the analysis for the velocity profile, the temperature profile is obtained as

$$\frac{T - T_o}{T_\infty - T_o} = \frac{3y}{\delta} - \frac{3y^2}{\delta^2} + \frac{y^3}{\delta^3} \qquad for \quad 0 \le y \le \delta \qquad (III.47a)$$

$$\frac{T - T_o}{T_\infty - T_o} = 1 \qquad\qquad\qquad for \quad y \ge \delta \qquad (III.47b)$$

while the velocity profile remains

$$u = u_\infty\left(\frac{3y}{\delta} - \frac{3y^2}{\delta^2} + \frac{y^3}{\delta^3}\right) \quad for \quad 0 \le y \le \delta$$

$$= u_\infty \qquad\qquad\qquad\qquad for \quad y \ge \delta \qquad from \quad (III.38)$$

215

The heat flux, q, is obtained, using the temperature profile, as

$$q = -k \frac{\partial T}{\partial y}\bigg|_{y=0} = \frac{3k}{\delta}(T_o - T_\infty) \qquad (III.48)$$

We can get an expression for the Nusselt number at length, x, as

$$Nu_x = \frac{q.x}{k\,\Delta T} = \frac{3k(T_o - T_\infty)x}{\delta k(T_o - T_\infty)} = \frac{3}{7.483}\mathrm{Re}_x^{\frac{1}{2}}.\mathrm{Pr}^{\frac{1}{2}} = 0.401\mathrm{Re}_x^{\frac{1}{2}}.\mathrm{Pr}^{\frac{1}{2}} \qquad (III.49)$$

Solution of the von Karman Integral Equations using the Displacement and Momentum Thickness Approach

A different approach, to simply using the velocity or temperature profiles to solve the von Karman equation, is to use the displacement and momentum thicknesses of the boundary layer (Kay, 1965) illustrated in the diagram below.

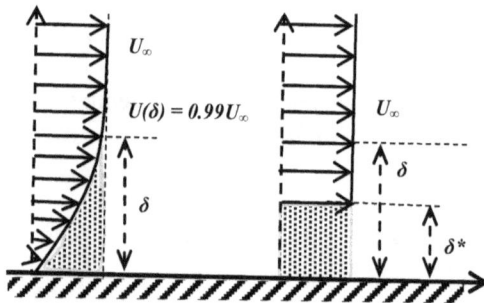

Displacement Thickness of a Boundary Layer

The shaded areas, within the boundary layer thickness and its equivalent displacement thickness, are equal.

In a hydrodynamic boundary layer of thickness, δ, the displacement thickness, δ^*, is defined as the distance that, when multiplied by the free stream velocity, equals the integral of velocity defect, $(u_\infty - u)$, across the boundary layer. It is a measure of the lost volume in boundary layer flow. That is,

$$\rho_\infty u_\infty \delta^* = \int_0^\delta (\rho_\infty u_\infty - \rho u)\,dy \qquad (III.50)$$

from which, for incompressible flow,

$$\delta^* = \int_0^{\delta} \left(1 - \frac{u}{u_\infty}\right) d y \qquad (III.51)$$

and for compressible flow,

$$\delta^* = \int_0^{\delta} \left(1 - \frac{\rho u}{\rho_\infty u_\infty}\right) d y \qquad (III.52)$$

The displacement thickness may, also, be defined as the reference distance in the fluid from the surface of the wall such that the flow per unit width, Q, from this distance to the edge of the boundary layer, at the free stream velocity, u_∞, is equal to the actual flow in the boundary layer. That is

$$Q = \int_0^{\delta} u\, d y = u_\infty \left(\delta - \delta^*\right) \qquad (III.53)$$

from which, for incompressible flow,

$$\delta^* = \int_0^{\delta} d y - \int_0^{\delta} \frac{u}{u_\infty} d y = \int_0^{\delta} \left(1 - \frac{u}{u_\infty}\right) d y \quad same\ as\ (III.51)$$

The displacement thickness is important in iterative boundary layer solutions. By successively using it in the boundary layer equations to calculate the displacement thickness along the wall, creating a new virtual wall and displacing the wall outward by the displacement thickness to obtain a new inviscid solution, slightly different free-stream conditions than the initial calculation are obtained. The boundary layer solution is then recalculated, using the new free-stream conditions, for the real wall. The process is repeated until the displacement thickness stops moving with each iteration.

The *momentum thickness*, θ, is defined as the distance that, when multiplied by the square of the free stream velocity, equals the integral of the momentum defect, $\rho u\left(u_\infty - u\right)$, across the boundary layer. It is a measure of lost momentum in boundary layer flow. That is,

$$Lost\ Momentum, M = \rho_\infty u_\infty^2 \theta = \int_0^{\delta} \rho u \left(u_\infty - u\right) d y \qquad (III.54)$$

From which, for incompressible flow,

$$\theta = \int_0^{\delta} \frac{u}{u_\infty}\left(1 - \frac{u}{u_\infty}\right) d y \qquad (III.55)$$

and for compressible flow,

217

$$\theta = \int_0^\delta \frac{\rho u}{\rho_\infty u_\infty}\left(1 - \frac{u}{u_\infty}\right)dy \qquad (III.56)$$

The *momentum thickness*, θ, may, also, be described as that distance from the surface of the wall between it and the edge of the boundary layer in which the momentum of the fluid, expressed in the free stream velocity, is equal to that in the boundary layer. Schematically, the situation may be visualised as shown below

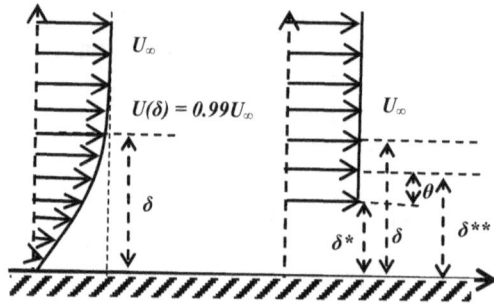

Momentum Thickness, θ, of a Boundary Layer

In this case,

$$Momentum\ in\ bl,\ M_{bl} = \int_0^\delta \rho u^2 dy = \rho_\infty u_\infty^2\left(\delta - \delta^{**}\right) \qquad (III.57)$$

from which, for incompressible flow,

$$\delta^{**} = \int_0^\delta \left(1 - \frac{u^2}{u_\infty^2}\right)dy \qquad (III.58)$$

The momentum thickness can be seen from the figure to be

$$\theta = \delta^{**} - \delta^* = \int_0^\delta \left[\left(1 - \frac{u^2}{u_\infty^2}\right) - \left(1 - \frac{u}{u_\infty}\right)\right]dy$$

$$= \int_0^\delta \frac{u}{u_\infty}\left(1 - \frac{u}{u_\infty}\right)dy \qquad (III.59)$$

while the momentum in the boundary layer is, now, seen to be

$$M_{bl} = \rho_\infty u_\infty^2\left(\delta - \delta^* - \theta\right) \qquad (III.60)$$

If we now consider the mass and momentum balance in a control volume ABCD over a small element of the the boundary layer, as shown below.

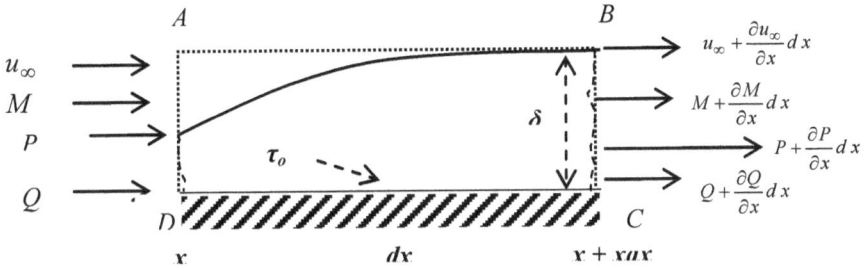

$$A \qquad\qquad\qquad\qquad B$$

$$u_\infty + \frac{\partial u_\infty}{\partial x} dx$$

$$u_\infty$$
$$M$$
$$P$$
$$M + \frac{\partial M}{\partial x} dx$$
$$\delta$$
$$P + \frac{\partial P}{\partial x} dx$$
$$Q$$
$$\tau_0$$
$$Q + \frac{\partial Q}{\partial x} dx$$

$$D \qquad\qquad\qquad\qquad C$$

$$x \qquad\qquad dx \qquad\qquad x + xax$$

Momentum Balance in a Boundary Layer

The net decrease in flow and momentum in the control volume over **dx** will be balanced by the pressure loss and surface friction forces. That is

$$Net\ rate\ of\ outflow of\ x - m om entum = -\rho u_\infty \frac{\partial Q}{\partial x} dx + \frac{\partial M}{\partial x} dx \quad (III.61)$$

$$Sum of\ shear and\ pressure\ forces = -\tau_0 dx - \delta. \frac{\partial P}{\partial x} dx \qquad (III.62)$$

Since these forces must be equal to each other

$$-\tau_0 dx - \delta. \frac{\partial P}{\partial x} dx = -\rho u_\infty \frac{\partial Q}{\partial x} dx + \frac{\partial M}{\partial x} dx$$

$$-\tau_0 - \delta. \frac{\partial P}{\partial x} = -\rho u_\infty \frac{\partial Q}{\partial x} + \frac{\partial M}{\partial x} \qquad (III.63)$$

is, now, another way of expressing the von Karman integral momentum equation. We can substitute the values for τ_0, P, Q and M from equations (III.33), (III.53) and (III.60) to get

$$-\tau_0 + \rho u_\infty \delta. \frac{\partial u_\infty}{\partial x} = -\rho u_\infty \frac{\partial}{\partial x} [u_\infty (\delta - \delta^*)] + \frac{\partial}{\partial x} [\rho u_\infty^2 (\delta - \delta^* - \theta)] \quad (III.64)$$

This can be simplified as shown below

$$-\tau_0 + \rho u_\infty \delta. \frac{\partial u_\infty}{\partial x} = -\rho u_\infty \left[u_\infty \frac{\partial (\delta - \delta^*)}{\partial x} + (\delta - \delta^*) \frac{\partial u_\infty}{\partial x} \right]$$

$$+ \rho u_\infty^2 \frac{\partial (\delta - \delta^* - \theta)}{\partial x} + (\delta - \delta^* - \theta) \frac{\partial \rho u_\infty^2}{\partial x}$$

$$= -\rho u_\infty \left[u_\infty \frac{\partial (\delta - \delta^*)}{\partial x} + (\delta - \delta^*) \frac{\partial u_\infty}{\partial x} \right]$$

$$+ \rho u_\infty^2 \frac{\partial (\delta - \delta^* - \theta)}{\partial x} + (\delta - \delta^* - \theta) \frac{\partial \rho u_\infty^2}{\partial u_\infty} \frac{\partial u_\infty}{\partial x}$$

$$= -\rho u_\infty^2 \frac{\partial(\delta - \delta^*)}{\partial x} - \rho u_\infty (\delta - \delta^*)\frac{\partial u_\infty}{\partial x}$$

$$+ \rho u_\infty^2 \frac{\partial(\delta - \delta^*)}{\partial x} - \rho u_\infty^2 \frac{\partial \theta}{\partial x} + 2\rho u_\infty (\delta - \delta^*)\frac{\partial u_\infty}{\partial x} - 2\rho u_\infty \theta \frac{\partial u_\infty}{\partial x}$$

to get the following form of the von Karman integral momentum equation

$$\frac{\tau_o}{\rho u_\infty^2} = \frac{\partial \theta}{\partial x} + \frac{(\delta^* + 2\theta)}{u_\infty}\frac{\partial u_\infty}{\partial x} \qquad (III.65)$$

We know that, outside the boundary layer, $\dfrac{\partial u_\infty}{\partial x} = 0$.

Hence from (III.65)

$$\frac{\tau_o}{\rho u_\infty^2} = \frac{\partial \theta}{\partial x} \qquad (III.66)$$

By assuming a velocity profile $u = f\,(y)$ and using $\tau_o = \mu \dfrac{\partial u}{\partial y}\bigg|_{y=0}$ a

solution of the momentum equation can be obtained.

For example, if the velocity profile is

$$u = u_\infty \left[\frac{3y}{\delta} - \frac{3y^2}{\delta^2} + \frac{y^3}{\delta^3}\right] \quad for \quad 0 \le y \le \delta \qquad (III.67)$$

$$\tau_o = \mu \frac{\partial u}{\partial y}\bigg|_{y=0} = \mu u_\infty \left[\frac{3}{\delta} - \frac{6y}{\delta^2} + \frac{3y^2}{\delta^3}\right]_{y=0} = \frac{3\mu u_\infty}{\delta} \qquad (III.68)$$

From (III.51) and (III.67)

$$\delta^* = \int_0^\delta \left(1 - \frac{3y}{\delta} + \frac{3y^2}{\delta^2} - \frac{y^3}{\delta^3}\right)dy = \left[y - \frac{3y^2}{2\delta} + \frac{y^3}{\delta^2} - \frac{y^4}{4\delta^3}\right]_0^\delta$$

$$= \delta - \frac{3\delta^2}{2\delta} + \frac{\delta^3}{\delta^2} - \frac{\delta^4}{4\delta^3} - 0 = \frac{\delta}{4} \qquad (III.69)$$

From (III.55) or (III.59) and (III.67)

$$\theta = \int_0^\delta \left(\frac{3y}{\delta} - \frac{3y^2}{\delta^2} + \frac{y^3}{\delta^3}\right)\left\{1 - \left[\frac{3y}{\delta} - \frac{3y^2}{\delta^2} + \frac{y^3}{\delta^3}\right]\right\}dy$$

$$= \int_0^\delta \left\{\left(\frac{3y}{\delta} - \frac{3y^2}{\delta^2} + \frac{y^3}{\delta^3}\right) - \left(\frac{3y}{\delta} - \frac{3y^2}{\delta^2} + \frac{y^3}{\delta^3}\right)\left(\frac{3y}{\delta} - \frac{3y^2}{\delta^2} + \frac{y^3}{\delta^3}\right)\right\}dy$$

$$= \int_0^\delta \left(\frac{3y}{\delta} - \frac{3y^2}{\delta^2} + \frac{y^3}{\delta^3} \right) d\,y - \int_0^\delta \left(\frac{9y^2}{\delta^2} - 9\left(\frac{y}{\delta}\right)^3 + 3\left(\frac{y}{\delta}\right)^4 \right) d\,y$$

$$+ \int_0^\delta \left(9\frac{y^3}{\delta^3} - 9\left(\frac{y}{\delta}\right)^4 + 3\left(\frac{y}{\delta}\right)^5 \right) d\,y - \int_0^\delta \left(3\frac{y^4}{\delta^4} - 3\left(\frac{y}{\delta}\right)^5 + \left(\frac{y}{\delta}\right)^6 \right) d\,y$$

$$= \left[\frac{3y^2}{2\delta} - \frac{y^3}{\delta^2} + \frac{y^4}{4\delta^3} \right]_0^\delta - \left[\frac{3y^3}{\delta^2} - \frac{9y^4}{4\delta^3} + \frac{3y^5}{5\delta^4} \right]_0^\delta$$

$$+ \left[\frac{9y^4}{4\delta^3} - \frac{9y^5}{5\delta^4} + \frac{y^6}{2\delta^5} \right]_0^\delta - \left[\frac{3y^5}{5\delta^4} - \frac{y^6}{2\delta^5} + \frac{y^7}{7\delta^6} \right]_0^\delta$$

$$= \frac{3}{2}\delta - \delta + \frac{\delta}{4} - 3\delta + \frac{9\delta}{4} - \frac{3\delta}{5} + \frac{9\delta}{4} - \frac{9\delta}{5} + \frac{\delta}{2} - \frac{3\delta}{5} + \frac{\delta}{2} - \frac{\delta}{7}$$

$$= \left(\frac{210-140+35-420+315-84+315-252+70-84+70-20}{140} \right)\delta$$

$$= \frac{15}{140}\delta = \frac{3}{28}\delta \qquad\qquad (III.70)$$

From equations (III.66), (III.68) and (III.70)

$$\frac{\tau_o}{\rho u_\infty^2} = \frac{\partial \theta}{\partial x} = \frac{3}{28}\frac{d\delta}{dx} = \frac{3\mu}{\rho u_\infty \delta} \qquad\qquad (III.71)$$

Or

$$\delta d\delta = \frac{28\mu}{\rho u_\infty}d\,x \qquad\qquad \text{as in } (III.41)$$

which, on integration and rearranging, gives

$$\frac{\delta}{x} = 7.483\left(\frac{\mu}{\rho u_\infty x}\right) = \frac{7.483}{\sqrt{Re_x}} \qquad\qquad \text{as in } (III.42)$$

A similar treatment for the thermal boundary layer by Kay (1965) gives the thermal thickness, ψ, as

$$\psi = \int_0^\delta \frac{u}{u_\infty}\left(1 - \frac{T}{T_\infty} \right) d\,y \qquad\qquad (III.72)$$

Using the same velocity distribution as in equation (III.67) and a temperature distribution as in equation (III.47a) given by

$$\frac{T-T_o}{T_\infty - T_o} = \frac{3y}{\delta} - \frac{3y^2}{\delta^2} + \frac{y^3}{\delta^3} \qquad\qquad \text{from } (III.47a)$$

$$\delta^* = \int_0^\delta \left(1 - \frac{u}{u_\infty}\right) dy = \int_0^\delta \left(1 - \frac{3y}{\delta} + \frac{3y^2}{\delta^2} - \frac{y^3}{\delta^3}\right) dy$$

$$= \left[y - \frac{3y^2}{2\delta} + \frac{y^3}{\delta^2} - \frac{y^4}{4\delta^3}\right]_0^\delta = \delta - \frac{3\delta}{2} + \delta - \frac{\delta}{4} = \frac{\delta}{4} \qquad (III.73)$$

$$\psi = \int_0^\delta \frac{u}{u_\infty}\left(1 - \frac{T}{T_\infty}\right) dy = \frac{3\delta}{28} \qquad from \quad (III.70)$$

The enthalpy, H, is given by

$$H = \rho u_\infty Cp\,(T_o - T_\infty)\!\left(\delta - \delta^* - \psi\right) = \rho u_\infty Cp\,(T_o - T_\infty)\!\left[\delta - \frac{\delta}{4} - \frac{3\delta}{28}\right]$$

$$= \frac{9}{14}\delta\rho u_\infty Cp(T_o - T_\infty) \qquad (III.74)$$

The heat flux, q, is obtained, using the temperature profile, as

$$q = -k\frac{\partial T}{\partial y}\bigg|_{y=0} = \frac{3k}{\delta}(T_o - T_\infty) \quad from \quad (III.48)$$

The energy equation is

$$-\frac{q}{\rho u_\infty Cp T_\infty} = \frac{d\psi}{dx} \qquad (III.75)$$

That is

$$-\frac{3k(T_\infty - T_o)}{\delta\,\rho u_\infty Cp(T_o - T_\infty)} = \frac{3d\delta}{28dx} \quad or \quad \delta d\delta = \frac{28k}{\rho u_\infty Cp}dx \qquad (III.76)$$

This gives

$$\frac{\delta^2}{2} = \frac{28kx}{\rho u_\infty Cp} \quad or \quad \frac{\delta}{x} = \frac{7.483}{Re_x^{\frac{1}{2}}.Pr^{\frac{1}{2}}} \qquad (III.77)$$

We can get an expression for the Nusselt number at length, x, as

$$Nu_x = \frac{q.x}{k\,\Delta T} = \frac{3k(T_o - T_\infty)x}{\delta k(T_o - T_\infty)} = \frac{3}{7.483}Re_x^{\frac{1}{2}}.Pr^{\frac{1}{2}} = 0.401 Re_x^{\frac{1}{2}}.Pr^{\frac{1}{2}} \qquad (III.78)$$

Solutions for other configurations may be obtained from Welty (1978), Kay (1965). Integral solutions give results not very different from exact solutions and yet are easier to apply and much more versatile in dealing with practical situations. It is second only to numerical methods as the methods of choice.

APPENDIX IV
Solutions by Dimensionless Analysis

Dimensionless analysis was initially applied to problems in which the mathematical relationships between the many variables of the system were not explicitly known. Increasingly, it was found to be an easier and faster way of determing working relationships between variables. It also makes possible the generalisation of experimental data and cuts down on experimentation since effects of groups of variables, not the individual variables, are now examined.

Dimensionless analysis may be defined as the mathematical tool by which the relationships between variables of a physical system are determined in a precise and dimensionless manner (Perry & Green, 1984). It is based on fundamental dimensions and is particularly useful where other methods of analysis are difficult or impossible to use.

The result of dimensionless analysis is an equation or groups of equations which show how one set of dimensionless groups are related to each other or to another set.

Dimensions and Units

The properties of the physical world are expressed in *dimensioned* form. The dimensions are expressed in *units*. Three systems of units are in common use.

i. The Engineering System

In this system, the primary and independent dimensions are *Mass, M; Length, L; Time, θ; Force, F*. But since, according to Newton's second law of motion

$$F = Mass \; x \; Acceleration = M.a \qquad (IV.1)$$

where, *a*, the acceleration, has the fundamental units of L/θ^2. *F*, consequently, will have the fundamental units of ML/θ^2.

To make this system dimensionally consistent with the definition of *F* as a fundamental dimension in this system, a constant, $1/g_C$ must multiply the acceleration. g_C will have to have the fundamental units of $ML/F\theta^2$ but will be, numerically, equal to the acceleration due to gravity at sea level, of 32. Thus, in this system

$$F = Mass\, x\, Acceleration = M\, x\, \frac{a}{g_C} \qquad (IV.2)$$

This was also known as the English system.

ii. The Gravitational System

Here, **Force, F, Length, L,** and **Time, θ**, are considered to be the primary dimensions having independent units. Mass is then a dependent dimension of **F, L** and **θ** with units defined, from (IV.1), by

$$Mass = \frac{Force}{Acceleration} = \frac{F\,\theta^2}{L} \qquad (IV.3)$$

No relating constant such as **g** is required, since Force, F, and Mass, M, are not considered independent, dimensionally. This was the American system.

iii. The Absolute System

Here **Mass, M, Length, L,** and **Time, θ**, are the primary dimensions with independent units. Force is then a dependent dimension with units ML/θ^2. No relating constant, **g**, is required because **F** and **M** are not independent. This was the French or European System. The S.I. system is based on the absolute system.

The various characteristics of all three systems are shown in the Tables IV.I and IV.2.

Dimensionless Groups

Dimensionless groups are unique and distinct combinations of variables, relevant in a physical situation which are dimensionless and represent the relationships between dominant variables which describe natural phenomena in the system. For example, the Reynolds number, in fluid mechanics, is dimensionless and is given by

$$Re = \frac{\rho D u}{\mu}, \frac{kg}{m^3}.\frac{m}{1}.\frac{m}{s}.\frac{m.s}{kg}$$

$$= \frac{\rho u}{\mu/D} \propto \frac{convective\,forces}{viscous\,forces} \qquad (IV.4)$$

Other dimensionless groups are similarly derived. The major

dimensionless groups in fluid mechanics, heat transfer, mass transfer and chemical reaction are listed in Tables IV.3 to IV.6.

Table IV.1:   Various Systems of Units in English Units (Streeter, 1971)

	English System		Gravitational System		Absolute System	
Quantity	**Dimension**	**Unit**	**Dimension**	**Unit**	**Dimension**	**Unit**
Time	θ or t	Second	θ or t	Second	θ or t	Second
Length	L	Foot	L	Foot	L	Foot
Mass	M	Pound mass	$F\theta^2/L$	Slug	M	Poundmass
Force	F	Pound force	F	Pound force	ML/θ^2	Poundal
Work	FL	Foot Pound Force	FL	Foot Pound force	ML^2/θ^2	Ft-Poundal
Temperature	T	Degree Fahrenheit absolute	T	Deg. F. abs. or Rankine Foot Pound Force	T	Deg. F. absolute or Rankine
Heat	Q	Btu.	FL	None	$\dfrac{ML^2}{\theta^2}$	Foot-Poundal
g_c	$\dfrac{ML}{F\theta^2}$	32.174 lb mass.ft per lb force.s^2	None	None	None	None
J.	$\dfrac{FL}{Q}$	778.26 ft.lbf per Btu	None		None	None

Table IV.2: Various Systems of Units in Metric Units (Streeter, 1971)

Quantity	English System		Gravitational System		Absolute System (S. I.)			
	Dimen-sion	Units	Dimen-sion	Units	Dimen-sion	Units	Dimension	Units
Time	θ	Seconds	θ	Seconds	θ	Seconds	θ	Seconds
Length	L	cm	L	cm	L	cm	L	meter
Mass	M	gram	$\frac{F\theta^2}{L}$	gram	M	gram	M	kg
Force	F	gram force	F	gram force	ML/θ^2	Dyne	$\frac{ML}{\theta^2}$	Newton
Work	FL	cm.gm. force	FL	cm.gm force	ML^2/θ^2	Erg	ML^2/θ^2	Joule
Temp	T	Deg C or K	T	Deg C or K	T	Deg C or K	T	Deg K
Heat	Q	Calories	FL	cm.gram	ML^2/θ^2	Erg	ML^2/θ^2	Joule
g_c	$ML/F\theta^2$	980.7 gmass.cm /gforce s²	None	None	None	None	None	None
J	$\frac{FL}{Q}$	42699 cm.gforce /Calorie	None	None	None	None	None	None

Note that g_c is not the same as g.

Table IV.3: Summary of Major Dimensionless Groups in Mass Transfer (Perry & Green, 1984)

Name	Symbol	Formula	Special Notes
1. Lewis Number	Le	$k/\rho C_p D_v$	D_v = molecular diffusivity
2. Peclet Number	Pe	Lu/D_1	D_1 = diffusion coefficient
3. Schmidt Number	Sc	$\mu/\rho D_v$	D_v = diffusion coefficient
4. Sherwood Number	Sh	$k_c L/D_v$	k_c = mass transfer coefficient
5. Rayleigh Number	Ra	$\dfrac{L^3 \rho^2 g\beta C_p \Delta T}{\mu k}$	ΔT = Temperature difference across film. β = expansion coefficient

Table IV.4: Summary of Major Dimensionless Groups in Heat Transfer (Perry & Green, 1984)

	Name	Symbol	Formula	Special Notes
1	Biot Number	Bi	hL/k	L = midpoint to surface
2.	Fourier Number	Fo	$\dfrac{kt}{\rho C_p L^2}$	k = thermal conductivity
3.	Graetz Number	Gz	WC_p/kL	W = mass flowrate
4.	Grashof Number	Gr	$L^3 \rho^2 g \beta \Delta T/\mu^2$	
5.	Nusselt Number	Nu	hD/k	D = diameter
6.	Peclet Number	Pe	$\dfrac{\rho u C_p L}{k}$	u = velocity
7.	Prandtl Number	Pr	$Cp\mu/k$	

Table IV.5: Summary of Major Dimensionless Groups in Chemical Reaction Engineering (Perry & Green, 1984)

	Name	Formula	Special Notes
1.	Arrhenius Group	$\Delta E/RT$	ΔE = energy change
2.	Damkohler Group I	$\dfrac{uL}{R_A C_A}$	C_A = concentration R_A = reaction rate
3.	Damkohler Group II	$\dfrac{uL^2}{D_v C_A}$	
4.	Damkohler Group III	$\dfrac{QuL}{\rho C_p V T}$	Q = heat generated T = Temperature
5.	Damkohler Group IV	$\dfrac{QuL^2}{kT}$	k = rate constant

**Table IV.6: Summary of Major Dimensionless Groups in Fluid Mechanics
(Perry & Green, 1984)**

	Name	Symbol	Formula	Special Notes
1.	Bingham Number	N_{Bm} or Bm	$\dfrac{\tau_y\,g_c\,L}{\mu_p V}$	τ_y = yield stress; μ_p = coefficient of rigidity
2.	Cauchy Number	N_c or C	$\rho V^2/E_b$	E_b = Bulk modulus
3.	Drag Coefficient	C_D	$(\rho_p-\rho_f)\,L.g/\rho_p u^2$	ρ_p = density of object; ρ_f = density of fluid
4.	Euler Number	Eu	$\dfrac{g_c(\Delta P_f/\rho)}{u^2}$	ΔP_f; ρ = friction heat
5.	Fanning Friction Factor	f	$\dfrac{g_c D(\Delta P_f/\rho)}{2u^2 L}$	D = pipe diameter; L = length of pipe
6.	Froude Number	Fr, N_{Fr}	u^2/gL	
7.	Knudsen Number	Kn, N_{Kn}	λ/L	λ = mean free path
8.	Mach Number	Ma,	u/u_c	u_c = velocity of sound in fluid
9.	Reynold's Number	Re, N_{Re}	$\dfrac{\rho D u}{\mu}$	
10.	Heat Capacity ratio	γ	Cp/Cv	
11.	Weber Number	We, N_{We}	$\rho L V^2/g_c\sigma$	σ = surface tension

Methods of Dimensionless Analysis

1. Rayleigh's Method of Indices

If n quantities Q_1..... Q_n are involved in a physical system, their mutual dependence can be expressed as

$$Q_1 = k Q_2^{a_2} . Q_3^{a_3} ..... Q_n^{a_n} \qquad (IV.5)$$

where k is a constant, a_i are indices and the Q_i include all variables and dimensional constants involved in the system such that the equation is dimensionally homogenous.

An equation is dimensionally homogenous if each of the primary dimensions on each side of the equal sign is raised to the same power. Specifically, if there are n variables and r primary dimensions, there will be a minimum $n - r$ dimensionless groups and $n - r - 1$ unrestricted indices, related as above, such that these indices make the expression dimensionally homogenous.

Consider the case of determining the pressure drop when a fluid flows through a pipe of internal diameter D and length L. The variables expected to be involved in the system are:

	Variable	Primary Dimensions
1	Pressure drop, ΔP	$\dfrac{Force}{Area} \approx \dfrac{M L}{\theta^2} . \dfrac{1}{L^2} = \dfrac{M}{L\theta^2}$
2	Pipe internal diameter, D	Length $\approx$ L
3	Pipe length, L	Length $\approx$ L
4	Pipe roughness, ε	Length $\approx$ L
5	Fluid velocity, u	$\dfrac{Length}{Time} \approx \dfrac{L}{\theta} = \dfrac{L}{\theta}$
6	Fluid absolute viscosity, μ	$\dfrac{Force}{Area/Time} \approx \dfrac{M L}{\theta^2} . \dfrac{\theta}{L^2} = \dfrac{M}{L\theta}$
7	Fluid density, ρ	$\dfrac{Mass}{Volume} \approx \dfrac{M}{L^3}$

Number of Variables, **n** = 7
Number of independent dimensions, **r** = 3 (M, L, θ)

Expected number of dimensionless groups, $n - r = 7\text{-}3 = 4$
Number of unrestricted indices, $n - r - 1 = 7 - 3 - 1 = 3$.
Hence the pressure drop, ΔP, will be expected to relate to the other variables as

$$\Delta P = k\, L^a\, D^b\, \varepsilon^c\, u^d\, \mu^e\, \rho^f \qquad (IV.6)$$

Expressed in fundamental dimensions, equation (IV.6) reduces to

$$\frac{M}{L\theta^2} = k\, L^a\, L^b\, L^c \left(\frac{L}{\theta}\right)^d \left(\frac{M}{L\theta}\right)^e \left(\frac{M}{L^3}\right)^f \qquad (IV.7)$$

Hence, for dimensional homogeneity (**k** is a constant of proportionality and is dimensionless), we must have

for   Mass, M $1 = e + f$ (IV.8)

for   Length, L $-1 = a + b + c + d - e - 3f$ (IV.9)

for   Time, θ $-2 = -d - e$ (IV.10)

If we choose the three unrestricted indices to be **a, c** and **e**, then, we must express **b, d** and **f** in terms of these unrestricted indices. Thus

from (IV.8) $f = 1 - e$ (IV.11)

from (IV.10) $d = 2 - e$ (IV.12)

from (IV.9), (IV.11) and (IV.12)

$$-1 = a + b + c + 2 - e - e - 3(1 - e) \quad or \quad b = -a - c - e \qquad (IV.13)$$

Hence, using (IV.6), the four dimensionless groups are obtained as follows:

$$\Delta P = k\, L^a\, D^{-a-c-e}\, \varepsilon^c\, u^{2-e}\, \mu^e\, \rho^{1-e} = k\,\rho u^2 \left(\frac{L}{D}\right)^a \cdot \left(\frac{\varepsilon}{D}\right)^c \cdot \left(\frac{\mu}{\rho u D}\right)^e$$

or $$\frac{\Delta P}{\rho u^2} = k \left(\frac{L}{D}\right)^a \cdot \left(\frac{\varepsilon}{D}\right)^c \cdot \left(\frac{\mu}{\rho u D}\right)^e \qquad (IV.14)$$

The first term on the left hand side of equation (IV.14) is the Euler number, the second and third terms on the right hand side are, respectively, the length to diameter ratio and the surface roughness to diameter ratio. The last term on the right hand side is the inverse of the Reynolds number.

The constants, **k, a, c** and **e**, are determined from experiments. Note, also, that different forms of dimensionless groups will result if different unrestricted exponents are selected.

2. Buckingham II - Theorem

If an equation is dimensionally homogenous, it can be reduced to a relationship among a complete set of dimensionless products. A set of dimensionless products of given variables is said to be complete if

    a. each product in the set is independent of the others and
    b. every other dimensionless product of the variables is a product of the powers of dimensionless products in the set. (Perry & Green, 1984).

Thus for a system of n physical variables or quantities and m independent variables, there will be p dimensionless groups constituting a complete set for the n quantities such that $p = n - m$ and

$$\pi_1 = \phi_1^{a_1} \cdot \phi_2^{a_2} \cdot \phi_m^{a_m} \cdot \phi_{m+1}$$

$$\pi_2 = \phi_1^{a_1} \cdot \phi_2^{a_2} \cdot \phi_m^{a_m} \cdot \phi_{m+2}$$

$$\cdots\cdots\cdots$$

$$\pi_p = \phi_1^{a_1} \cdot \phi_2^{a_2} \cdot \phi_m^{a_m} \cdot \phi_n \qquad (IV.15)$$

where $\varphi_1 \ldots \varphi_m$ do not form a dimensionless group. The p dimensional groups are related by the general equation.

$$\phi\left(\pi_1, \pi_2, \pi_3, \pi_4, \cdots \pi_p, \right) = 0 \qquad (IV.16)$$

To illustrate the use of the method, consider the heat transfer between a pipe and a fluid flowing through it. The variables in the system are:

	Variable	Primary Dimensions
1	Film heat transfer coefficient, h	$\dfrac{Heat\,Energy}{Area \times Time \times Temp\,Difference} \approx \dfrac{M\,L^2}{\theta^2} \cdot \dfrac{1}{L^2} \cdot \dfrac{1}{\theta} \cdot \dfrac{1}{T} = \dfrac{M}{\theta^3 \cdot T}$
2	Internal Dia, D	Length $\approx L = L$
3	Fluid Velocity, u	$\dfrac{Length}{Time} \approx \dfrac{L}{\theta} = \dfrac{L}{\theta}$
4	Fluid Density, ρ	$\dfrac{Mass}{Volume} \approx \dfrac{M}{L^3}$
5	Fluid absolute Viscosity, μ	$\dfrac{Force}{Area/Time} \approx \dfrac{M\,L}{\theta^2} \cdot \dfrac{\theta}{L^2} = \dfrac{M}{L\theta}$

6	Fluid Thermal Conductivity, k	$\dfrac{Heat\,Energy}{Area\,x\,Tim\,ex\,Temp\,Gradient} \approx \dfrac{M\,L^2}{\theta^2}\cdot\dfrac{1}{L^2}\cdot\dfrac{1}{\theta}\cdot\dfrac{L}{T}$
		$= \dfrac{M\,.L}{\theta^3\,.T}$
7	Fluid Specific Heat, Cp	$\dfrac{Heat\,Energy}{Mass\,x\,Temp\,Difference} \approx \dfrac{M\,L^2}{\theta^2}\cdot\dfrac{1}{M}\cdot\dfrac{1}{T} = \dfrac{L^2}{\theta^2\,.T}$

Number of variables, $n = 7$
Number of independent dimensions, $r = 4$ (M, L, θ, T)
Hence number of dimensionless groups, $n - r = 3$

If we choose D, u, μ, and k as the independent dimensions, that is, those which do not form a dimensionless group, then, h, Cp and ρ will form dimensionless groups. Thus, from (15)

$$\pi_1 = D^a.u^b.\,\mu^c.k^d.h$$
$$\pi_2 = D^a.u^b.\,\mu^c.k^d.Cp$$
$$\pi_p = D^a.u^b.\,\mu^c.k^d.\rho \qquad\qquad (IV.17)$$

In terms of the fundamental dimensions, equation (IV.17) becomes

$$\pi_1 = D^a.u^b.\,\mu^c.k^d.h = L^a.\left(\dfrac{L}{\theta}\right)^b.\left(\dfrac{M}{L.\theta}\right)^c.\left(\dfrac{M.L}{\theta^3.T}\right)^d.\left(\dfrac{M}{\theta^3.T}\right)$$

$$\pi_2 = D^a.u^b.\,\mu^c.k^d.Cp = L^a.\left(\dfrac{L}{\theta}\right)^b.\left(\dfrac{M}{L.\theta}\right)^c.\left(\dfrac{M.L}{\theta^3.T}\right)^d.\left(\dfrac{L^2}{\theta^2.T}\right)$$

$$\pi_p = D^a.u^b.\,\mu^c.k^d.\rho = L^a.\left(\dfrac{L}{\theta}\right)^b.\left(\dfrac{M}{L.\theta}\right)^c.\left(\dfrac{M.L}{\theta^3.T}\right)^d.\left(\dfrac{M}{L^3}\right) \qquad (IV.18)$$

From the principle of dimensional homogeneity, the indices on both sides of equation (IV.17) are equal. Since the indices on the πs are zero, we get, for π_1 in equation (18), that

for	Mass, M	$0 = c + d + 1$	(IV.19)
for	Length, L	$0 = a + b - c + d$	(IV.20)
for	Time, θ	$0 = -b - c - 3d - 3$	(IV.21)
For	Temperature, T	$0 = -d - 1$	(IV.22)

232

From which

$$d = -1 \qquad \text{(IV.23)}$$

$$c = 0 \qquad \text{(IV.24)}$$

$$b = 0 \qquad \text{(IV.25)}$$

$$a = 1 \qquad \text{(IV.26)}$$

From (IV.23), (IV.24), (IV.25), (IV.26) and (IV.17)

$$\pi_1 = D^1.u^0.\,\mu^0.k^{-1}.h = \frac{hD}{k}, \text{ the Nusselt number} \qquad \text{(IV.27)}$$

Similarly

$$\pi_2 = D^0.u^0.\,\mu^1.k^{-1}.Cp = \frac{Cp\,\mu}{k}, \text{ the Prandtl number} \qquad \text{(IV.28)}$$

$$\pi_3 = D^1.u^1.\,\mu^{-1}.k^0.\rho = \frac{\rho u D}{\mu}, \text{ the Reynolds number} \qquad \text{(IV.29)}$$

Hence by Buckingham's π Theorem, equation (IV.16)

$$\phi\left(\pi_1, \pi_2, \pi_3, \pi_4, \cdots \pi_p, \right) = 0 \quad or \quad \phi(Nu, \text{Pr}, \text{Re}) = 0 \qquad \text{(IV.30)}$$

One version of equation (IV.30) in popular use is

$$Nu = \phi(\text{Pr}, \text{Re}) \qquad \text{(IV.31)}$$

Experimentally, it is found that equation (IV.31) is of the explicit form

$$Nu = k.\text{Pr}^a.\text{Re}^b \qquad \text{(IV.32)}$$

where **k, a, b** are determined from experiments using

$$\log Nu = \log k + a\log \text{Pr} + b\log \text{Re} \qquad \text{(IV.33)}$$

APPENDIX V
Solutions from Mass, Momentum and Heat Transfer Analogies

The basis of the method is to compare the given situation to the other modes of transfer namely mass, heat and momentum. In heat transfer, for example, an analogy is sought to an equivalent situation in momentum transfer for which the governing equations are known. Desired parameters in heat transfer are then evaluated from what are known in momentum transfer. Similarly, mass transfer analogies to heat and momentum transfer can be made.

There are four major analogies. All of them deal with the estimation of either the heat transfer coefficient (for heat transfer), the mass transfer coefficient (for mass transfer) or the friction factor (for momentum transfer). All are derived essentially from mixing length theories in momentum transfer. However, only how these analogies are used in heat and momentum transfer will be discussed in detail here. The relevant analogies are;

i.   The Reynold's Analogy

ii.   The Prandtl Analogy, also known as the Taylor - Prandtl Analogy

iii.   The von - Karman Analogy

iv.   The Colburn Analogy

1. The Reynold's Analogy

This analogy may be better visualised if we considered heat and momentum transfer very close to a heated wall in temperature equilibrium. The Reynold's analogy assumes that fluid flow is turbulent right up to the wall so that heat and momentum transfer is by convection. That is, there is no layer of stagnant fluid near the wall.

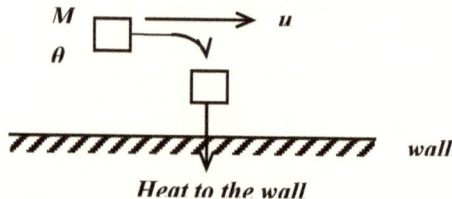

Heat to the wall

Consider a small element of fluid of mass, **M**, at a temperature, θ, above that of the wall, travelling with velocity, **u**. In time, **t**, it travels to the wall and transfers heat to it. If **R** is the shear stress it generates at the surface of the wall of area, **A**, as a result of transfer of fluid of mass, **M**, then a momentum and heat flux balance may be obtained as follows.

Change in momentum per unit time
$$\frac{M\,u}{t} = R.A \qquad (V.1)$$

Heat flux
$$q = \frac{M\,Cp\,\theta}{t.A} = h.\theta \qquad (V.2)$$

Cp is the heat capacity of the fluid and **h** is the heat transfer coefficient

From equation (V.1)
$$\frac{M}{t.A} = \frac{R}{u} \qquad (V.3)$$

From equation (V.2)
$$\frac{M}{t.A} = \frac{h}{Cp} \qquad (V.4)$$

From (V.3) and (V.4) and dividing both sides by ρu, where ρ is the fluid density

$$\frac{h}{\rho u Cp} = \frac{R}{\rho u^2} \qquad (V.5)$$

$\dfrac{h}{\rho u Cp}$ is the Stanton Number and $\dfrac{R}{\rho u^2}$ is half the friction factor, **f**, which we know, from fluid mechanics, to be related to the Reynold's number. Thus

$$St = \frac{h}{\rho u Cp} = \frac{R}{\rho u^2} = \alpha \, Re^n \qquad (V.6)$$

where a and n are constants.

To determine the value of R, consider the laminar layer to be a Newtonian fluid in which the shear stress is given by

$$R = \mu \frac{d\,u}{d\,y} \qquad (V.7)$$

Because the layer is very thin,

$$\frac{d\,u}{d\,y} \cong \frac{u}{\delta} \qquad (V.8)$$

where δ is of the order of magnitude of the boundary layer dimensions. Thus, from (V.7) and (V.8) shear stress at the wall R_o is given by

$$R_o = \mu \frac{d\,u}{d\,y}\bigg|_{y=0} = \frac{\mu.u}{\delta} \qquad (V.9)$$

From (V.5) and (V.9)

$$\frac{h}{\rho u Cp} = \frac{R_o}{\rho u^2} = \frac{\mu . u}{\delta \rho u^2} \quad or \quad \frac{h}{Cp} = \frac{\mu}{\delta} \qquad (V.10)$$

Since $h = k/\delta$ where k is the thermal conductivity

$$\frac{h}{Cp} = \frac{k}{\delta Cp} = \frac{\mu}{\delta} \quad or \quad \frac{Cp\,\mu}{k} = Pr = \frac{\delta}{\delta} = 1 \qquad (V.11)$$

We can conclude from this that the Reynold's analogy applies only when $Pr = 1$ and when there is virtually no boundary layer.

2. The Prandtl Analogy or The Taylor-Prandtl Analogy

The physical configuration of the boundary layer assumed in the Reynold's analogy is idealized. In practice, there is a boundary layer as well as a turbulent flow region and each region is associated with clearly defined velocities, temperatures, etc. This may be represented, schematically, as shown below

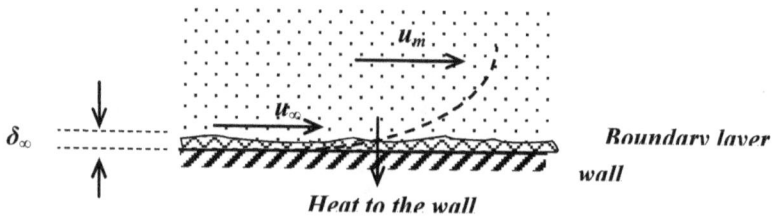

The Prandtl analogy assumes that fluid flow in the boundary layer is laminar. Thus, in the laminar layer:

Shear Stress at the wall
$$R_o = \mu \frac{du}{dy}\bigg|_{y=0} = \frac{\mu u_\infty}{\delta_\infty} \qquad (V.12)$$

Heat flux to the wall
$$q = -k \frac{dT}{dy}\bigg|_{y=0} = -k \frac{(T_\infty - T_o)}{\delta_\infty} \qquad (V.13)$$

Dividing (V.13) by (V.12) and substituting for $k = \dfrac{Cp\,\mu}{Pr}$

$$\frac{q}{R_o} = -\frac{k(T_\infty - T_o)}{\mu u_\infty} = -\frac{Cp(T_\infty - T_o)}{Pr u_\infty} \qquad (V.14)$$

Similarly, in the turbulent zone, to a good approximation

$$\frac{q}{R_o} = -\frac{Cp(T_m - T_\infty)}{(u_m - u_\infty)} \qquad (V.15)$$

From equation (V.14)

236

$$T_\infty = T_o - \frac{q\,Pr\,u_\infty}{Cp\,R_o} \tag{V.16}$$

Putting (V.16) in (V.15) and rearranging

$$\frac{q(u_m - u_\infty)}{Cp\,R_o} = T_\infty - T_m = T_o - \frac{q\,Pr\,u_\infty}{Cp\,R_o} - T_m = T_o - T_m - \frac{q\,Pr\,u_\infty}{Cp\,R_o}$$

That is $\quad \dfrac{q}{R_o}\left[u_m - u_\infty + Pr\,u_\infty\right] = \dfrac{q\,u_m}{R_o}\left[1 + \dfrac{u_\infty}{u_m}(Pr-1)\right] = Cp\left(T_o - T_m\right)$

From which $\quad \dfrac{q}{Cp\left(T_o - T_m\right)} = \dfrac{R_o}{u_m\left[1 + \dfrac{u_\infty}{u_m}(Pr-1)\right]} \tag{V.17}$

Dividing both sides by ρu_m

$$\frac{q}{\rho u_m\,Cp\left(T_o - T_m\right)} = \frac{R_o}{\rho u_m^2}\cdot\frac{1}{\left[1 + \dfrac{u_\infty}{u_m}(Pr-1)\right]} \tag{V.18}$$

But $\quad \dfrac{R_o}{\rho u_m^2} = \dfrac{f}{2}$ where f is the friction factor and $\dfrac{q}{\rho u_m\,Cp\left(T_o - T_m\right)} = St$

Hence, from (V.18)

$$St = \frac{f}{2}\cdot\frac{1}{\left[1 + \dfrac{u_\infty}{u_m}(Pr-1)\right]} \tag{V.19}$$

If $Pr = 1$, it reduces to the Reynold's analogy. For laminar flow in smooth pipes $\dfrac{f}{2} = 0.332 Re^{-\frac{1}{2}}$. Empirical data (Kay, 1965) indicate that $\dfrac{u_\infty}{u_m} = 1.5\,Re^{-\frac{1}{8}}.Pr^{-\frac{1}{6}}$. Substituting these into equation (V.19)

$$St = \frac{0.332 Re^{-\frac{1}{2}}}{\left[1 + 1.5\,Re^{-\frac{1}{8}}.Pr^{-\frac{1}{6}}(Pr-1)\right]} \tag{V.20}$$

Equation (V.20) is valid for systems for which $0.5 < Pr < 30$. Other versions of this equation are given in Table (V.1) below. Before using them, the student is advised to ascertain what ratio of the friction factor the C_f represents.

3. The von - Karman Analogy

The assumptions of the Reynold's analogy that there is no boundary layer or by the Taylor - Prandtl analogy that there is a sharp transition from the boundary layer to the turbulent zone do not represent reality. The von - Karman's analogy postulates that it is more realistic to expect a smooth transition of velocity or temperature from the wall through the laminar sublayer, through the buffer zone to the turbulent zone as shown in the schematic diagram below.

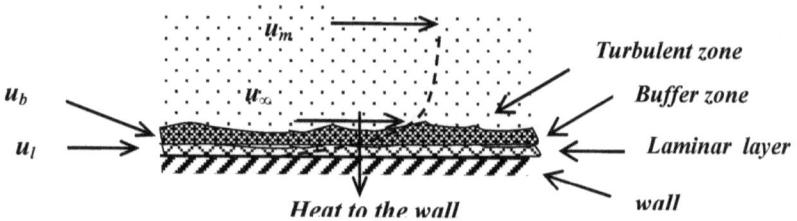

Corresponding temperature variables, T_m, T_∞, T_b and T_l, have not been inserted into the diagram to avoid clutter. The dimensionless distance, y^*, and velocity, u^* are first defined as

$$u^* = \frac{u}{\sqrt{\frac{\tau_0}{\rho}}} \quad \text{and} \quad y^* = \frac{y}{v}\sqrt{\frac{\tau_0}{\rho}} \tag{V.21}$$

where v is the momentum diffusivity. These are used to solve the Navier-Stokes equations to obtain the so called universal velocity profiles by means of which it becomes possible, among other things, to define the limits of the laminar sublayer, the buffer zone and the turbulent zone as shown in Table (V.2). The momentum and heat transfer in each zone are then determined as follows

i. <u>Laminar sublayer, $y^* < 5$ and $u^* = y^*$</u>

   <u>Momentum transfer</u>

   $$\tau_0 = \mu\frac{du}{dy} = \frac{\mu}{v}\sqrt{\frac{\tau_0}{\rho}}.\sqrt{\frac{\tau_0}{\rho}}.\frac{du^*}{dy^*} = \frac{\mu}{v}.\frac{\tau_0}{\rho}.\frac{du^*}{dy^*} = \frac{\mu\tau_0}{v\rho} \tag{V.22}$$

   determined solely by the shear stress at the wall since $u^* = y^*$.

   <u>Heat transfer</u>

   $$q = -k\frac{dT}{dy} = -\frac{k.\sqrt{\frac{\tau_0}{\rho}}}{v}\frac{dT}{dy^*} \quad \text{or} \quad dT = -\frac{qv}{k.\sqrt{\frac{\tau_0}{\rho}}}dy^* \tag{V.23}$$

238

which on integration: $\displaystyle\int_{T_o}^{T_l} dT = -\dfrac{qv}{k\cdot\sqrt{\dfrac{\tau_o}{\rho}}}\int_0^5 dy^*$ gives

$$T_l - T_o = -\dfrac{5qv}{k\cdot\sqrt{\dfrac{\tau_o}{\rho}}} = -\dfrac{5\,Pr}{\sqrt{\dfrac{\tau_o}{\rho}}}\cdot\dfrac{q}{\rho Cp} \qquad (V.24)$$

ii. Buffer Zone, $5 < y^* < 30$ and $u^* = -3.05 + 5 \ln y^*$

Within this zone, $\qquad u^* = -3.05 + 5\ln y^* \qquad$ (V.25)

Momentum transfer

$$\tau_o = \rho(v + \varepsilon)\dfrac{du}{dy} \qquad (V.26)$$

ε is the eddy diffusivity. Using the dimensionless definitions of (V.21) in (V.26)

$$\tau_o = \dfrac{\rho(v+\varepsilon)}{v}\cdot\sqrt{\dfrac{\tau_o}{\rho}}\sqrt{\dfrac{\tau_o}{\rho}}\cdot\dfrac{du^*}{dy^*} = \dfrac{\tau_o(v+\varepsilon)}{v}\cdot\dfrac{du^*}{dy^*} \qquad (V.27)$$

That is

$$1 = \dfrac{(v+\varepsilon)}{v}\cdot\dfrac{du^*}{dy^*} \qquad (V.28)$$

From (V.25)

$$\dfrac{du^*}{dy^*} = \dfrac{5}{y^*} \qquad (V.29)$$

Putting (V.29) in (V.28) and solving for ε,

$$1 = \dfrac{5(v+\varepsilon)}{vy^*} \quad or \quad \varepsilon = v\left(\dfrac{y^*}{5} - 1\right) \qquad (V.30)$$

At the inner edge of the buffer zone, $y^* = u_l^* = 5$

$$u_l^* = -3.05 + 5\ln 5 \qquad (V.31)$$

At the outer edge of the buffer zone, $y^* = 30$

$$u_b^* = -3.05 + 5\ln 30 \qquad (V.32)$$

Hence

$$u_b^* = u_l^* + 5\ln 6 = 5(1 + \ln 6) \qquad (V.33)$$

And from (V.21)

$$u_b = u_b^* . \sqrt{\frac{\tau_o}{\rho}} = 5 \sqrt{\frac{\tau_o}{\rho}} (1 + \ln 6) \qquad (V.34)$$

Heat transfer

$$q = - \rho Cp (\alpha + \varepsilon) \frac{dT}{dy} \qquad (V.35)$$

α is the thermal diffusivity.

Using the dimensionless definitions of (V.21) and using the value of ε in (V.35)

$$q = - \rho Cp \left[\alpha + v \left(\frac{y^*}{5} - 1 \right) \right] \frac{\sqrt{\frac{\tau_o}{\rho}}}{v} \frac{dT}{dy^*}$$

$$= - \rho Cp \left[\alpha - v + v \frac{y^*}{5} \right] \frac{\sqrt{\frac{\tau_o}{\rho}}}{v} \frac{dT}{dy^*}$$

which, on rearrangement, becomes

$$\int_{T_I}^{T_b} dT = - \frac{q}{\rho Cp} . \frac{v}{\sqrt{\frac{\tau_o}{\rho}}} . \int_5^{30} \frac{dy^*}{\left(\alpha - v + v \frac{y^*}{5} \right)}$$

$$= - \frac{q}{\rho Cp} . \frac{5v}{\sqrt{\frac{\tau_o}{\rho}}} . \int_5^{30} \frac{dy^*}{\alpha \left(5 - \frac{5v}{\alpha} + \frac{v}{\alpha} y^* \right)}$$

and integration gives

$$T_b - T_I = - \frac{q}{\rho Cp} . \frac{5v}{\alpha \, Pr \sqrt{\frac{\tau_o}{\rho}}} \ln [5 - 5 Pr + Pr y^*]_5^{30}$$

$$= - \frac{q}{\rho Cp} . \frac{5}{\sqrt{\frac{\tau_o}{\rho}}} \ln \left(\frac{5 + 25 Pr}{5} \right)$$

That is

$$T_b - T_I = - \frac{q}{\rho Cp} . \frac{5}{\sqrt{\frac{\tau_o}{\rho}}} \ln (1 + 5 Pr) \qquad (V.36)$$

240

iii. Turbulent Zone, $y* > 30$

Reynolds analogy applies

That is
$$\frac{q}{\tau_o} = -\frac{Cp\,(T_m - T_b)}{(u_m - u_b)} \qquad from \quad (V.15)$$

Thus
$$T_m - T_b = -\frac{q}{\tau_o\,Cp}\left(u_m - u_b\right) \qquad (V.37)$$

Substituting for u_b from equation (V.34) into (V.37)

$$T_m - T_b = -\frac{q}{\tau_o\,Cp}\left(u_m - 5\sqrt{\frac{\tau_o}{\rho}}\,(1 + \ln 6)\right) \quad (V.38)$$

Since

$$f = \frac{\tau_o}{\frac{1}{2}\rho u_m^2} \quad or \quad \sqrt{\frac{\tau_o}{\rho}} = u_m\sqrt{\frac{f}{2}} \qquad (V.39)$$

$$T_m - T_b = -\frac{q}{\tau_o\,Cp}\left(u_m - 5u_m\sqrt{\frac{f}{2}}.(1 + \ln 6)\right)$$

$$= -\frac{q\,u_m}{\rho Cp\left(\dfrac{\tau_o}{\rho}\right)}\left(1 - 5\sqrt{\frac{f}{2}}.(1 + \ln 6)\right)$$

$$= -\frac{q}{\rho Cp.\sqrt{\dfrac{f}{2}}.\sqrt{\dfrac{\tau_o}{\rho}}}\left[1 - 5\sqrt{\frac{f}{2}}.(1 + \ln 6)\right] \qquad (V.40)$$

We can add equations (V.24), (V.36) and (V.40) to obtain the temperature difference from the wall through the laminar and buffer zones to the turbulent zone to get

$$T_l - T_o + T_b - T_l + T_m - T_b = -\frac{5\,Pr}{\sqrt{\dfrac{\tau_o}{\rho}}}.\frac{q}{\rho Cp} - \frac{q}{\rho Cp}.\frac{5}{\sqrt{\dfrac{\tau_o}{\rho}}}\ln(1 + 5\,Pr)$$

$$- \frac{q}{\rho Cp.\sqrt{\dfrac{f}{2}}.\sqrt{\dfrac{\tau_o}{\rho}}}\left[1 - 5\sqrt{\frac{f}{2}}.(1 + \ln 6)\right]$$

That is

$$-T_o + T_m = -\frac{q}{\rho Cp\sqrt{\dfrac{\tau_o}{\rho}}}\left[5\,Pr + 5\ln(1+5\,Pr) + \frac{1}{\sqrt{\dfrac{f}{2}}}\left[1-5\sqrt{\frac{f}{2}}.(1+\ln 6)\right]\right]$$

$$=-\frac{q}{\rho Cp\sqrt{\dfrac{\tau_o}{\rho}}}\left[\frac{\sqrt{\dfrac{f}{2}}[5\,Pr + 5\ln(1+5\,Pr)]+\left[1-5\sqrt{\dfrac{f}{2}}.(1+\ln 6)\right]}{\sqrt{\dfrac{f}{2}}}\right]$$

That is

$$T_o - T_m = \Delta T = \frac{q}{\rho Cp\sqrt{\dfrac{f}{2}}.\sqrt{\dfrac{\tau_o}{\rho}}}\left[5\sqrt{\frac{f}{2}}.[(Pr-1)+\ln(1+5\,Pr)-\ln 6]+1\right] \quad (V.41)$$

The Stanton number is defined as

$$St = \frac{q}{\rho u_m\, Cp\,\Delta T} = \frac{q\sqrt{\dfrac{f}{2}}}{\rho\sqrt{\dfrac{\tau_o}{\rho}}Cp\,\Delta T} \qquad (V.42)$$

Substituting for $\dfrac{q}{\rho Cp\,\Delta T}$ from (V.41) into (V.42)

$$St = \frac{q}{\rho u_m\, Cp\,\Delta T} = \frac{q\sqrt{\dfrac{f}{2}}}{\rho\sqrt{\dfrac{\tau_o}{\rho}}Cp\,\Delta T}$$

$$=\frac{\sqrt{\dfrac{f}{2}}}{\sqrt{\dfrac{\tau_o}{\rho}}}.\frac{\sqrt{\dfrac{f}{2}}.\sqrt{\dfrac{\tau_o}{\rho}}}{5\sqrt{\dfrac{f}{2}}.[(Pr-1)+\ln(1+5\,Pr)-\ln 6]+1}$$

$$=\frac{\dfrac{f}{2}}{5\sqrt{\dfrac{f}{2}}.[(Pr-1)+\ln(1+5\,Pr)-\ln 6]+1} \qquad (V.43)$$

242

4. **The Chilton-Colburn Analogy**

The Chilton-Colburn analogy, also known as the j-factor analogy, is said to be probably the most successful and widely used analogy for heat, momentum, and mass transfer. It is based on the similarity of mechanism and mathematics among the transport processes of momentum, heat and mass transfer.

In momentum and heat transfer in a fluid moving on a flat plate, it is found that, at any distance, x, along the plate

$$Nu_x = \alpha \, \mathrm{Re}_x^{\frac{1}{2}} \, \mathrm{Pr}^{\frac{1}{3}} \qquad (V.44)$$

where α is a constant. If we divide both sides of (V.44) by $\mathrm{Re}_x \, \mathrm{Pr}^{1/3}$, we find that

$$\frac{Nu_x}{\mathrm{Re}_x \, \mathrm{Pr}^{\frac{1}{3}}} = \alpha \, \mathrm{Re}_x^{-\frac{1}{2}} = \frac{f}{2} \qquad (V.45)$$

Multiplying the numerator and denominator of the LHS of equation (V.45)) by $\mathrm{Pr}^{2/3}$, we get

$$\frac{Nu_x}{\mathrm{Re}_x \, \mathrm{Pr}^{\frac{1}{3}}} \frac{\mathrm{Pr}^{\frac{2}{3}}}{\mathrm{Pr}^{\frac{2}{3}}} = \frac{Nu_x . \mathrm{Pr}^{\frac{2}{3}}}{\mathrm{Re}_x \, \mathrm{Pr}} = \alpha \, \mathrm{Re}_x^{-\frac{1}{2}} = \frac{f}{2} \qquad (V.46)$$

But

$$\frac{Nu_x}{\mathrm{Re}_x \, \mathrm{Pr}} = \frac{h.x}{k} . \frac{\mu}{\rho.x.u} . \frac{k}{Cp.\mu} = \frac{h}{\rho.u.Cp} = St \qquad (V.47)$$

Thus we can conclude that

$$j_H = St.\mathrm{Pr}^{\frac{2}{3}} = \alpha \, \mathrm{Re}_x^{-\frac{1}{2}} = \frac{f}{2} = j_M \qquad (V.48)$$

where j_H and j_M are the j-factors for heat and for momentum transfer, respectively.

A similar analysis for mass transfer yields

$$j_D = \frac{k_C}{v} . Sc^{\frac{2}{3}} = j_H = St.\mathrm{Pr}^{\frac{2}{3}} = \frac{\alpha}{\mathrm{Re}_x^{\frac{1}{2}}} = \frac{f}{2} = j_M \qquad (V.49)$$

where j_D is the j-factor for mass transfer. k_C is the mass transfer coefficient based on concentration.

The Chilton-Colburn analogy is valid for fully developed turbulent flow in conduits with $Re > 10000$, $0.7 < Pr < 160$, and in tubes where $L/d > 60$ It is less complex than the von - Karman and Taylor - Prandtl analogies while being more accurate than the Reynold's analogy.

A wider range of data is said to be correlated by the Friend-Metzner analogy (Price, 2010) which gives

$$Nu_b = \frac{Re_b . Pr_b . \frac{f}{2} . \left(\frac{\mu_b}{\mu_w}\right)^{0.14}}{1.20 + 11.8. \sqrt{\frac{f}{2}} (Pr_b - 1). Pr_b^{-\frac{1}{3}}} \qquad (V.50)$$

for heat and momentum transfer, and

$$\frac{Sh}{Re.Sc} = \frac{\frac{f}{2}}{1.20 + 11.8. \sqrt{\frac{f}{2}} (Sc_b - 1). Sc_b^{-\frac{1}{3}}} \qquad (V.51)$$

for momentum and mass transfer. These expressions are valid for Re > 10000, 0.5 < Pr < 600, 0.5 < Sc < 3000.

APPENDIX VI
Solutions by Numerical Methods

Numerical Solutions

Numerical solutions of mathematical equations or expressions find application where explicit or implicit solutions are either not possible or tedious. Here, we shall be concerned only with the solution of differential equations. The general principle is to introduce, into the physical or mathematical system, an appropriate set of independent co-ordinate values such as $x_i = x_o + i\Delta x$, $y_i = y_o + j\Delta y$, $z_i = z_o + k\Delta z$, (where $i, j, k = 0, \pm1, \pm2, ...$etc.) on the basis of which the dependent variables such as $u(x,y,z)$, $T(x,y,z)$, etc., can be expressed.

For first order linear differential equations describing initial value problems, the appropriate numerical methods are the Euler method or its modified form, Heun's formula, and the Runge-Kutta algorithms. For second order linear differential equations describing boundary value problems, the appropriate method is the use of finite difference approximations. These were introduced in chapter one of volume one of this book in connection with the solution of problems in heat conduction.

Basic Definitions of Finite Difference Approximations

Finite difference algebra is a full subject in its own right and cannot be fully treated here. We can, however, summarise its aspects relevant to numerical analaysis of heat convection problems.

Finite Differences

The finite difference is the discrete analog of the derivative. Only three forms are commonly considered: forward, backward, and central differences. The first forward difference of the function, U(x), is defined as

$$\Delta U(x) = U(x+h) - U(x) \qquad (VI.1)$$

where h is a very small non-zero increment in the value of x. Depending on the application, h may be variable or held constant.

A backward difference uses the function values at x and $x - h$, instead of the values at $x + h$ and x. That is

$$\nabla U(x) = U(x) - U(x - h) \qquad (VI.2)$$

The central difference is given by

$$\delta U(x) = U(x + \tfrac{1}{2}h) - U(x - \tfrac{1}{2}h) \qquad (VI.3)$$

To illustrate how finite differences can be used, consider the function $U(x) = x^3$. We can tabulate the first, second, third and fourth forward differences in a difference table as follows:

x	x^3	Δ	Δ^2	Δ^3	Δ^4
1	1				
		7			
2	8		12		
		19		6	
3	27		18		0
		37		6	
4	64		24		
		61			
5	125				

It can be seen that the third forward difference is constant at 6. This idea is, sometimes used to detect values that are in error in a table of experimental data. Similar tables can be constructed for the backward and central differences.

The next step is to be convinced that the continuous derivatives of our differential equations can be, satisfactorily, approximated by discrete quantities used in finite difference mathematics.

Recall that the basic definition of the derivative of a continuously differentiable function U(x), in calculus, is

$$\frac{dU(x)}{dx} = \lim_{\Delta x \to o} \left(\frac{U(x + \Delta x) - U(x)}{\Delta x} \right) \qquad (VI.4)$$

If, instead of Δx tending to zero, we make it infinitessimally small, finite and equal to a value, say h, we can try to approximate the continuous differential by the finite difference ratio

$$\frac{dU(x)}{dx} \approx \frac{U(x + h) - U(x)}{h} \qquad (VI.5)$$

Hence, the first forward difference divided by h approximates the derivative when h is small, with an error of the order of magnitude of h. The first backward difference, also, has the same magnitude of error. The central difference, however, has an error of the order, h^2, and is, theoretically, therefore, more accurate although it has stability problems

246

in use.

We, also, know from the Taylor series around a point, x_i, that

$$U(x_i + h) = U(x_i) + \frac{h}{1!}\left(\frac{dU(x)}{dx}\right)_i + \frac{h^2}{2!}\left(\frac{d^2U(x)}{dx^2}\right)_i + \frac{h^3}{3!}\left(\frac{d^3U(x)}{dx^3}\right)_i + \cdots (VI.6)$$

Similarly

$$U(x_i - h) = U(x_i) - \frac{h}{1!}\left(\frac{dU(x)}{dx}\right)_i + \frac{h^2}{2!}\left(\frac{d^2U(x)}{dx^2}\right)_i - \frac{h^3}{3!}\left(\frac{d^3U(x)}{dx^3}\right)_i + \cdots (VI.7)$$

If in (VI.6) we neglect terms of order 2 and higher

$$U(x_i + h) = U(x_i) + \frac{h}{1!}\left(\frac{dU(x)}{dx}\right)_i \quad or \quad \frac{dU(x)}{dx} = \frac{U(x_i + h) - U(x_i)}{h} \quad see\ (VI.5)$$

the first forward finite difference representation of the derivative. If in (VI.7) we neglect terms of order 2 and higher

$$U(x_i - h) = U(x_i) - \frac{h}{1!}\left(\frac{dU(x)}{dx}\right)_i \quad or \quad \frac{dU(x)}{dx} = \frac{U(x_i) - U(x_i - h)}{h} \quad (VI.8)$$

the first backward finite difference representation of the derivative. If we add (VI.6) and (VI.7) and neglect terms of order 3 and higher, we get

$$\frac{d^2U(x)}{dx^2} = \frac{U(x_i + h) + U(x_i - h) - 2U(x_i)}{h^2} \quad (VI.9)$$

with an error of order h^3. If we subtract (VI.7) from (VI.6) and neglect terms of order 3 and higher, we get

$$\frac{dU(x)}{dx} = \frac{U(x_i + h) - U(x_i - h)}{2h} \quad (VI.10)$$

which is a first finite central difference representation of the derivative, with an error of order h.

Finite difference algebra has developed relationships, using the Taylor series, between the first forward difference, represented by Δ, the continuous derivative, represented by D and the backward difference represented by ∇ as follows

$$\Delta = hD + \frac{1}{2!}h^2D^2 + \frac{1}{3!}h^3D^3 + \cdots = e^{hD} - 1 \quad (VI.11)$$

This is equivalent to

$$hD = \ln(1 + \Delta) = \Delta - \frac{1}{2!}\Delta^2 + \frac{1}{3!}\Delta^3 - \cdots \quad (VI.12)$$

The analogous expressions for the backward and central difference operators
are, respectively

$$hD = -\ln(1-\nabla) \quad and \quad hD = 2\operatorname{arcsinh}\left(\tfrac{1}{2}\delta\right) \tag{VI.13}$$

Application to the Solution of Differential Equations

Consider a general second order linear differential equation

$$\frac{d^2 U(x)}{dx^2} + P(x)\frac{dU(x)}{dx} + Q(x)U(x) = f(x) \tag{VI.14}$$

with boundary conditions $U(a) = \alpha$ and $U(b) = \beta$ and $a = x_o < x_1 < x_2 < x_3 < \ldots \ldots < x_{n-1} < x_n = b$.

We can obtain the finite difference approximation of (VI.14) using (VI.8) and (VI.9) as

$$f(x) = \frac{U(x_i + h) + U(x_i - h) - 2U(x_i)}{h^2} + P(x)\frac{U(x_i) - U(x_i - h)}{h}$$

$$+ Q(x)U(x) \tag{VI.15}$$

We can simplify the notation as shown in the Table below

Original Notation	Simplified Notation
$U(x_i)$	U_i
$U(x_i + h)$	U_{i+1}
$U(x_i - h)$	U_{i-1}
$f(x_i)$	f_i
$P(x_i)$	P_i
$Q(x_i)$	Q_i

This enables us to get a simplified version of (VI.15) as

$$h^2 f_i = \left(1 + \frac{h}{2}P_i\right)U_{i+1} + \left(h^2 Q_i - 2\right)U_i + \left(1 - \frac{h}{2}P_i\right)U_{i-1} \tag{VI.16}$$

If we let i take on values 1, 2, 3......n-1, we get $n-1$ equations in the $n-1$ unknowns, U_1, U_2, U_3,.....U_{n-1} since we know U_o and U_n already from the boundary conditions. These $n-1$ equations can then be solved by any of the many methods of matrix algebra.

It is to be noted that if the differential equation to be solved is nonlinear, it has to be linearised in some fashion.

References for Appendices I to VI

1   Kay J. M.; An Introduction to Fluid Mechanics and Heat Transfer; 2nd edn; Cambridge University Press, London, 1965

2   Korn G. A., Korn T. M.; Mathematical Handbook For Scientists and Engineers, 2nd edn.; McGraw-Hill Book Co., N.Y., U.S.A., 1968

3   http:/en.wikipedia.org/wiki/Derivation_of_the_Navier%E2%80%93Stokes_equations, 20/11/2009

4   Juster G.L.Y; Chuma.cas.usf.edu/justerGLY5932/Notespart1

5   Perry R., Green D.; Chemical Engineers' Handbook, 6th. edn.; McGraw - Hill Book Co.,, N. Y., U.S.A., 1984

6   Sargent, R. W. H., Class Notes, Imperial College, London, 1966

7   Boundary Layer Parameters.
www.aerojockey.com/papers/bl/node2.html.

8   Price R. M.; Determination of Mass Transfer Coefficients; www.cbu.edu/~rprice/lectures/mtcoeff.html

9   Weisstein, Eric W.; "Finite Difference." From *MathWorld*--A Wolfram Web Resource.
http://mathworld.wolfram.com/FiniteDifference.html

10   Zill D. G., Cullen M. R.; Differential Equations with Boundary Value Problems, 4th edn; Chapter 9; Brooks/Cole Publishing Company; Ca, USA, 1997

APPENDIX VII: Tables of Common Properties of Materials

Table VII.1a: Common Properties of Water, H_2O
(www.engineeringtoolbox.com)

Molecular Weight, 18.02
Triple Pt. Temp, 273.2 K

Normal Boiling Point 373.2 K
Critical Point, 647.3 K

t, C	Absolute pressure, P, kN/m²	Density, ρ, kg/m³	Specific volume, v, 10^{-3}, m³/kg	Dynamic Viscosity μ, centipoise	Kinematic viscosity, v, 10^{-6}, m²/s
0 (Ice)		916.8			
0.01	0.6	999.8	1.00	1.78	
4 (maximum density)	0.9	1000.0			
5	0.9	1000.0	1.00	1.52	
10	1.2	999.8	1.00	1.31	1.304
15	1.7	999.2	1.00	1.14	
20	2.3	998.3	1.00	1.00	1.004
25	3.2	997.1	1.00	0.890	
30	4.3	995.7	1.00	0.798	0.801
35	5.6	994.1	1.01	0.719	
40	7.7	992.3	1.01	0.653	0.658
45	9.6	990.2	1.01	0.596	
50	12.5	988	1.01	0.547	0.553
55	15.7	986	1.01	0.504	
60	20.0	983	1.02	0.467	0.474
65	25.0	980	1.02	0.434	
70	31.3	978	1.02	0.404	0.413
75	38.6	975	1.03	0.378	
80	47.5	972	1.03	0.355	0.365
85	57.8	968	1.03	0.334	
90	70.0	965	1.04	0.314	0.326
95	84.5	962	1.04	0.297	
100	101.33	958	1.04	0.281	0.295

t ,C	Absolute pressure, P, kN/m^2	Density, ρ, kg/m^3	Specific volume, v, 10^{-3}, m^3/kg	Dynamic Viscosity μ, centipoise	Kinematic viscosity, v, 10^{-6}, m^2/s
105	121	954	1.05	0.267	
110	143	951	1.05	0.253	
115	169	947	1.06	0.241	
120	199	943	1.06	0.230	0.249
125	228	939	1.06	0.221	
130	270	935	1.07	0.212	
135	313	931	1.07	0.204	
140	361	926	1.08	0.196	0.215
145	416	922	1.08	0.190	
150	477	918	1.09	0.185	
155	543	912	1.10	0.180	
160	618	907	1.10	0.174	0.189
165	701	902	1.11	0.169	
170	792	897	1.11	0.163	
175	890	893	1.12	0.158	
180	1000	887	1.13	0.153	0.170
185	1120	882	1.13	0.149	
190	1260	876	1.14	0.145	
195	1400	870	1.15	0.141	
200	1550	864	1.16	0.138	0.158
220		840			0.149
225	2550	834	1.20	0.121	
240		814			0.142
250	3990	799	125	0.110	
260		784			0.137
275	5950	756	1.32	0.0972	
300	8600	714	1.40	0.0897	
325	12130	654	1.53	0.0790	
350	16540	575	1.74	0.0648	
360	18680	528	1.90	0.0582	

Table VII.1b: Common Properties of Water, H_2O (contd)
(www.engineeringtoolbox.com)

t, C	Absolute pressure, P, kN/m^2	Specific Heat, Cp, kJ/kgK	Specific enthalpy, H, kJ/kg	Expansion coefficient, β, 10^{-3} (1/K)	Prandtl's Number
0.01	0.6	4.210	0	-0.07	13.67
5	0.9	4.204	21.0	0.160	
10	1.2	4.193	41.9	0.088	9.47
15	1.7	4.186	62.9	0.151	
20	2.3	4.183	83.8	0.207	7.01
25	3.2	4.181	104.8	0.257	
30	4.3	4.179	125.7	0.303	5.43
35	5.6	4.178	146.7	0.345	
40	7.7	4.179	167.6	0.385	4.34
45	9.6	4.181	188.6	0,420	
50	12.5	4.182	209.6	0.457	3.56
55	15.7	4.183	230.5	0.486	
60	20.0	4.185	251.5	0.523	2.99
65	25.0	4.188	272.4	0.544	
70	31.3	4.191	293.4	0.585	2.56
75	38.6	4.194	314.3	0.596	
80	47.5	4.198	335.3	0.643	2.23
85	57.8	4.203	356.2	0.644	
90	70.0	4.208	377.2	0.665	1.96
95	84.5	4.213	398.1	0.687	
100	101.33	4.219	419.1	0.752	1.75
105	121	4.226	440.2		
110	143	4.233	461.3		
115	169	4.240	482.5		
120	199	4.248	503.7	0.860	1.45
125	228	4.26	524.3		
130	270	4.27	546.3		
135	313	4.28	567.7		

t, C	Absolute pressure, P, kN/m^2	Specific Heat, Cp, kJ/kgK	Specific enthalpy, H, kJ/kg	Expansion coefficient, β, 10^{-3} (l/K)	Prandtl's Number
140	361	4.29	588.7	0.975	1.25
145	416	4.30	610.0		
150	477	4.32	631.8		
155	543	4.34	653.8		
160	618	4.35	674.5	1.098	1.09
165	701	4.36	697.3		
170	792	4.38	718.1		
175	890	4.39	739.8		
180	1000	4.42	763.1	1.233	0.98
185	1120	4.45	785.3		
190	1260	4.46	807.5		
195	1400		829.9		
200	1550	4.51	851.7	1.392	0.92
220		4.63		1.597	0.88
225	2550	4.65	966.8		
240		4.78		1.862	0.87
250	3990	4.87	1087		
260		4.98		2.21	0.87
275	5950	5.20	1211		
300	8600	5.65	1345		
325	12130	6.86	1494		
350	16540	10.1	1672		
360	18680	14.6	1764		

Table VII.2: Thermal Properties of Steam, H_2O
(www.engineeringtoolbox.com)

T , K	Density, g/cc	Heat Capacity, J/g.K	Thermal Conductivity, W/cm.K	Viscosity, g/cm.sec	Prandtl Number, Pr
2000	1.09E-04	2.832	2.06E-04	6.27E-05	0.86
1500	1.46E-04	2.594	1.48E-04	4.91E-05	0.86
1000	2.20E-04	2.267	8.80E-05	3.43E-05	0.88
800	2.75E-04	2.130	6.60E-05	2.81E-05	0.91
600	3.66E-04	2.003	4.60E-05	2.14E-05	0.93
500	4.41E-04	1.947	3.65E-05	1.77E-05	0.94
400	5.55E-04	1.900	2.77E-05	1.40E-05	0.96
373.2	5.98E-04	2.020	2.48E-04	1.20E-04	0.98
370	5.38E-04	2.020	2.46E-04	1.19E-04	0.98
360	3.78E-04	1.980	2.37E-04	1.15E-04	0.96
350	2.60E-04	1.950	2.30E-04	1.11E-04	0.94
340	1.74E-04	1.930	2.23E-04	1.07E-04	0.93
330	1.14E-04	1.910	2.17E-04	1.03E-04	0.91
320	7.15E-05	1.890	2.10E-04	9.89E-05	0.89
310	4.36E-05	1.880	2.04E-04	9.49E-05	0.87
300	2.55E-05	1.870	1.98E-04	9.09E-05	0.86
290	1.42E-05	1.860	1.92E-04	8.69E-05	0.84
280	7.60E-06	1.850	1.86E-04	8.29E-05	0.82
273.2	4.80E-06	1.850	1.82E-04	7.94E-05	0.81

- T(K) = Temperature in Kelvin, t (C) = Temperature in Celsius
- t (C) = T (K) - 273
- Heat Capacity [=] 1 J/gK = 0.23864 BTU/lbm/F
- Thermal Conductivity [=] 1 W/cmK = 57.77 BTU/hr/ft/F
- Viscosity [=] 1 g/cm sec = 1 poise = 241.9 lbm/ft/hr

Table VII.3: Common Properties of Air
(www.engineeringtoolbox.com)

Temperature, t, ^oC	Density ρ, kg/m^3	Specific heat capacity, Cp, $kJ/kg\,K$	Thermal conductivity, k, $W/m\,K$	Kinematic viscosity, v, $(m^2/s) \times 10^{-6}$	Expansion coefficient, β, $(1/K) \times 10^{-3}$	Prandtl's number, Pr
-150	2.793	1.026	0.0116	3.08	8.21	0.76
-100	1.980	1.009	0.0160	5.95	5.82	0.74
-50	1.534	1.005	0.0204	9.55	4.51	0.725
0	1.293	1.005	0.0243	13.30	3.67	0.715
20	1.205	1.005	0.0257	15.11	3.43	0.713
40	1.127	1.005	0.0271	16.97	3.20	0.711
60	1.067	1.009	0.0285	18.90	3.00	0.709
80	1.000	1.009	0.0299	20.94	2.83	0.708
100	0.946	1.009	0.0314	23.06	2.68	0.703
120	0.898	1.013	0.0328	25.23	2.55	0.70
140	0.854	1.013	0.0343	27.55	2.43	0.695
160	0.815	1.017	0.0358	29.85	2.32	0.69
180	0.779	1.022	0.0372	32.29	2.21	0.69
200	0.746	1.026	0.0386	34.63	2.11	0.685
250	0.675	1.034	0.0421	41.17	1.91	0.68
300	0.616	1.047	0.0454	47.85	1.75	0.68
350	0.566	1.055	0.0485	55.05	1.61	0.68
400	0.524	1.068	0.0515	62.53	1.49	0.68

Table VII.4: Phase Change Properties of some Common Materials
(www.engineeringtoolbox.com)

Substance	Latent Heat Fusion, kJ/kg	Melting Point, °C	Latent Heat Vaporization, kJ/kg	Boiling Point, °C
Alcohol, ethyl	108	-114	855	78.3
Ammonia	339	-75	1369	-33.34
Carbon dioxide	184	-78	574	-57
Helium			21	-268.93
Hydrogen	58	-259	455	-253
Lead	24.5	327.5	871	1750
Nitrogen	25.7	-210	200	-196
Oxygen	13.9	-219	213	-183
R134a		-101	215.9	-26.6
Toluene		-93	351	110.6
Turpentine			293	
Water	334	0	2260 (at 100 C)	100

- $1\ m^2/s = 10^4\ St(Stokes) = 10^6\ cSt = 10.764\ ft^2/s = 38750\ ft^2/h$
- $1\ J/kg = 4.299x10^{-4}\ Btu/lb_m = 2.388x10^{-4}\ kcal/kg$

INDEX

259

www.ingramcontent.com/pod-product-compliance
Lightning Source LLC
Chambersburg PA
CBHW031807190326
41518CB00006B/231